T0221083

Stability and Trim
of
Fishing Vessels

Stability and Trim

of

Fishing Vessels

and other Small Ships

Second Edition

J. Anthony Hind, C.Eng., F.R.I.N.A.

Fishing News Books Ltd
Farnham · Surrey · England

First published 1967
Second edition 1982

British Library CIP data

Hind, J. Anthony
Stability and trim of fishing vessels
and other small ships. - 2nd ed.
1. Fishing boats - Design and construction - Safety measures
I. Title
623.8'22   VM431

ISBN: 9780852381212

Contents

Symbols and Abbreviations

	Abbreviation	*Symbol*
Acceleration due to gravity		*g*
After perpendicular	A.P.	*AP*
Amidships, between perpendiculars	—	*B* ⊗ *P*
Amidships, on load waterplane	—	*L* ⊗ *L*
Angle	—	θ
Area, in general	—	*A*
Area, midship section	—	*Am*
Area, waterplane	—	*Aw*
Between perpendiculars	B.P.	—
Breadth, in general	—	*B* or *b*
Breath, moulded	B.mld. or BM	*B*
Breadth, extreme	B. extr.	*B*
Centre of buoyancy	CB	*B*
Centre of buoyancy above moulded base line	—	*KB*
Centre of buoyancy, longitudinal	*LCB*	—
Centre of buoyancy, vertical	*VCB*	—
Centre of flotation	CF	*F*
Centre of flotation, longitudinal	*LCF*	—
Centre of gravity	CG	*G* or *g*
Centre of gravity above moulded base line	—	*KG*
Centre of gravity, longitudinal	LCG	—
Centre of gravity, vertical	VCG	—
Coefficient, block	—	*Cb*
Coefficient, midship section area	—	*Cm*
Coefficient, prismatic	—	*Cp*
Coefficient, waterplane area	—	*Cw*
Density	—	δ or *d*
Depth, moulded	D.mld or DM	*D*
Depth or distance in general	—	*d*
Diameter	Dia.	*D* or *d*
Displacement, weight of	—	*W* or Δ

	Abbreviation	*Symbol*
Displacement, volume of	—	V or ∇
Draft, moulded	dr. mld.	d
Draft, extreme (bottom of keel)	dr. extr. or dr. b.k.	d
Fore perpendicular	F.P.	FP
Freeboard		f
Height, centre of buoyancy above moulded base line	—	KB
Height, centre of gravity above moulded base line	—	KG
Height, metacentre above moulded base line	—	KM
Height, metacentre above centre of buoyancy (metacentric radius)	—	BM
Height, metacentric	—	GM
Inertia, moment of	—	I or i
Length between perpendiculars	LBP or Lpp	—
Length of ship in general	—	L
Length on load waterline	LWL	—
Length of ship overall	LOA	—
Metacentre	—	M
Metacentre above centre of buoyancy (metacentric radius)	—	BM
Metacentre above moulded base line	—	KM
Metacentric height	—	GM
Midship sectional area	—	Am
Moment to change trim $1''$	MCT $1''$	—
Origin	—	O
Permeability	—	μ
Pressure	—	p
Righting lever	GZ	GZ
Sum	—	Σ
Tons per inch immersion	TPI	—
Volume of displacement	—	V or ∇
Waterplane area	—	Aw
Weight	—	w

Some Greek Letters used in Technical Terms

Δ	δ	delta
Θ	θ	theta (usually denotes angles)
M	μ	mu
Π	π	pi (always = 3.142)
Σ	σ	sigma (Σ is the summation sign)

The symbol $\int$ (like an elongated s) is the integral sign used in calculus.

Ⓧ is a symbol peculiar to ships and shipbuilding, the origin of which is obscure. It always means 'Amidships'. It is frequently drawn on plans with an arrow vertically through the centre, in which case it indicates the *midship point*.

⅏   denotes the centreline

EXAMINATION OF SKIPPERS AND SECOND HANDS OF FISHING BOATS IN SHIP STABILITY

Chapter 8 Syllabuses

General

In each paper throughout the syllabuses, questions may be set combining one or more paragraphs.

The syllabus for a 'Full' certificate is always to be regarded as including the syllabus for a 'Limited' certificate of the same grade and the 'Full' certificate of a lower grade (if any).

With effect from 1 May 1967 certain changes were made in the syllabuses for Skippers and Second Hands to include questions on ship stability. These changes which were detailed in Department of Trade *Merchant Shipping Notice No. M. 512*, are now incorporated in the syllabuses.

SKIPPER (FULL and LIMITED)

Paper (2 hours)

Ship Stability

The examination in stability will be designed to show that the candidate has a working knowledge of the factors governing the stability of a fishing vessel. A candidate will be required to have a knowledge of:

(1) Centre of gravity, centre of buoyancy, metacentre and metacentric height.

(2) Stable, neutral and unstable equilibrium. Stiff and tender ships. Angle of loll, its cause and correction.

(3) Statical stability, stability curves and range of stability.

(4) The effect on the statical stability of raising or lowering of weights and the taking on board or discharging of weights such as fuel oil and fish. The effect of weights suspended at a height such as when the cod end is hoisted inboard. The effect of ice accretion on the upper works. Practical examples, requiring simple calculations, of the conditions in this sub-paragraph.

(5) The effect of freeboard on the range of stability.

(6) The effect of free surface on the metacentric height. Simple exercises in applying to the GM the correction for free surface

and to determine the effect on the centre of gravity due to adding or taking out weights.

(7) A knowledge of the stability information supplied to a fishing vessel.

Chapter 5   Prescribed Marking in the Examination

To pass in the written portion, a candidate will be required to obtain the appropriate percentage pass in the subjects shown in the following tables and also to obtain 70 per cent of the total marks for all subjects. The time and marks allotted for each written part of the examination for each grade of certificate will be as follows:

Second Hand (Limited)

First day	*Time*	*Total marks*	*Percentage pass mark*
1. Chartwork and pilotage	3 hrs.	200	70
2. Practical navigation	2 hrs.	200	70

Second or subsequent days
Orals.

Skipper (Limited)

First day	*Time*	*Total marks*	*Percentage pass mark*
1. Chartwork and pilotage	3 hrs.	200	70
2. Practical navigation I	2 hrs.	150	70
Second day			
3. Practical navigation II	3 hrs.	200	70
4. Ship stability	2 hrs.	100	50

Third or subsequent days
Orals.

Second Hand (Full)

First day	*Time*	*Total marks*	*Percentage pass mark*
1. Chartwork and pilotage	3 hrs.	200	70
2. Practical navigation I	2 hrs.	150	70
Second day			
3. Practical navigation II	3 hrs.	200	70
4. Principles of navigation	2 hrs.	150	—

Third or subsequent days
Orals.

Skipper (Full)

First day	Time	Total marks	Percentage pass mark
1. Practical navigation I	3 hrs.	200	70
2. Practical navigation II	2 hrs.	150	70

Second day

3. Chartwork and pilotage	3 hrs.	200	70
4. Navigational aids (including electronic aids)	2 hrs.	150	50

Third day

5. Ship stability	2 hrs.	100	50

Third or subsequent days
Orals.

Second Hand (Special)

First day	Time	Total marks	Percentage pass mark
1. Chartwork and pilotage	3 hrs.	See below	See below
2. Practical navigation	2 hrs.	See below	See below

Second or subsequent days
Orals.

Second Hand (Special)
(Modified examination)
Orals only.

SECOND HAND (SPECIAL, LIMITED, FULL)

Candidates for Certificates of Competency will, in the oral parts of the examination, in addition to other subjects, be required to understand and give satisfactory answers on the following:

General understanding of centre of gravity, centre of buoyancy and metacentric height and the effect of adding or removing weights. An understanding of the effect of weights suspended at a height such as when the cod end is hoisted inboard; the effect of free surface in fuel and ballast tanks and the effect of ice accretion on the upper works.

This oral section of the syllabuses for Second Hands' certificates means that candidates for Skippers' certificates are also liable to oral examination on points detailed above.

(Reproduced by permission of the Department of Trade.)

ACKNOWLEDGEMENTS

THE AUTHOR'S thanks are due to the late Mr. James L. Kent, C.B.E., a Vice-President of The Royal Institution of Naval Architects, for his perusal of the manuscript and for his several valuable suggestions which have been incorporated. Some of the recommendations of the International Maritime Organisation (previously Inter-governmental Maritime Consultative Organisation) on the stability of fishing vessels have been quoted, most of which will be found in the Appendices. The Author is indebted to Mr. J. L. E. Jens, C.Eng., FRINA, Senior Deputy Director, Maritime Safety Division of IMO for his kindness in this connection.

Author's Note to Second Edition

Since this book was first published in 1967, much of its advice and many of the principles behind its recommendations have become officially adopted. Some of them are now enshrined in United Kingdom and International legislation particularly as regards the stability criteria discussed in Chapter 5 under 'Stability Standards and Minimum Stability'. As far as the United Kingdom is concerned, the attention of Owners, Superintendents and Masters is drawn towards the official 'Fishing Vessels Safety Provisions Rules, S.I. No. 330:1975' (HMSO). These set out requirements for hull construction, stability, structural fire protection and fire detection, means of escape, watertight integrity, boilers and machinery, pumping arrangements, electrical equipment and installations, crew protection (bulwarks, guard rails *etc*), and the carriage of charts and publications on U.K. registered fishing vessels of 12 metres in length and over. They also incorporate the regulations governing life-saving appliances and fire appliances which have previously been part of general merchant shipping regulations.

The Rules provide for fishing vessels of 12 metres Registered length and over to be surveyed for compliance with these Rules and the Radio Rules and a United Kingdom Fishing Vessel Certificate (valid for 4 years subject to intermediate inspection of the vessel carried out between 21 and 27 months from the date of issue of the Certificate) is issued to each vessel on completion of the survey. The Rules apply to both new and existing vessels, but exemptions from the requirements of particular Rules is granted to existing vessels where it is found unreasonable to require compliance, provided the vessel is found to be safe in operation.

Surveys of Classed vessels by surveyors to Lloyd's Register of Shipping or by Sea Fish Industry Authority surveyors for grant and loan purposes are accepted as partial surveys for the issue of Certificates. The balance of these surveys is carried out by D.o.T. surveyors.

In the matter of the carriage of 'Stability Information Aboard Ships', Chapter 10, all new vessels contracted for after 1st May, 1975 are required under U.K. regulations to carry full stability information. Existing vessels were required to comply generally with this provision over a phased period which lasted up to 1981 or as determined by the Department of Trade (e.g. fishing vessels could be purchased from outside the U.K. after 1981 and these would still be 'existing' vessels). The provisions and the phasing for existing vessels is necessarily somewhat complicated to take account of varying circumstances and there will be some concessions regarding the extent and form of the information for the smaller existing vessels. All new vessels, however, irrespective of size, must fully comply at the outset. Reference should be made to S.I. No. 330:1975.

J. ANTHONY HIND, C.Eng., F.R.I.N.A.

Preface

WE have finally come to the end of the easy way of doing things.

Economic pressures and higher standards of living have forced attention to the doing of things more efficiently, to greater cost effectiveness and higher industrial productivity.

Consequently, the tools of commerce—the 'hardware'—have become more specialised and complex. This puts a premium on knowledge and skill in every field of commercial activity and not least in the business of fishing.

Fishing is as scientific an industry as most and a good deal more so than many. It is no longer either adequate or realistic merely to learn by rule of thumb and, although there is still no substitute for practical experience at sea, there is now not sufficient time just to 'pick it up as one goes along'.

The greater degree of technical knowledge and competence required aboard a modern fishing vessel today is the inevitable consequence of all this and, one may soberly observe, is now reflected in the Department of Trade examinations for Certificates of Competency for senior fishermen.

This is a text book. It was written primarily to help fulfil the need of those who have to pass an examination. An effort has been made to give the student a basic knowledge of the subject whilst at the same time keeping explanations clear and simple. Mathematics have not been needlessly proliferated, but they cannot be avoided entirely in a subject the treatment of which is essentially mathematical. Facility with basic mathematical rules and operations should first be acquired, therefore, before attempting any serious study of this book. There is no easy royal road to the understanding of stability problems and over-simplification would not lead to anything which was worthwhile in the end.

Nevertheless, let it be emphasised that there is *no black magic* either. Some Greek notation is used because Greek letters have standardised meanings. They are merely symbols which are distinctive like road signs—not some voodoo mystique. As just about all technical books and reference material use this common 'language' there is no point in encouraging students to remain ignorant of a code dotted all over the technical highway. It is not 'scientific jargon' any more than the special terms used in the description of fishing gear are. Every activity has to have its own special terminology otherwise there would be endless confusion. The advantage becomes clearer with use and many people would agree that £ and $ for instance are better or more striking 'jargon' than L or D!

But the student may take heart from the comments of my friend and acknowledged authority on ship stability, the late Lloyd Woollard, M.A, R.C.N.C. an Hon. Vice-President of The Royal Institution of Naval Architects, 'The stability of ships is the only branch of Naval Architecture for which the theory is exact. There is no danger of this theory being upset later by experimental evidence. . . . A thousand years hence the theory of stability will be much the same as it is today, although, perhaps, the methods of its calculation will be improved; but the theory, by its very nature, cannot change.' Yet it is some justification for this book that he also said at the same time, 'One would have thought, therefore, that by now, or even before now, we should have reached some degree of finality; yet we still get a stream of highly interesting papers which give us fresh food for thought on this subject.'

Apart from the chore of passing examinations, it is hoped that this book will find a wider use. More is contained therein than will be required for formal examinations. The object has been to give the interested student an overall appreciation and a practical understanding of the stability behaviour of small ships rather than a 'cook book' for examinations. Perhaps marine superintendents of small ship fleets may find it useful reading and reference, and it is not beyond hope that a copy might be useful aboard ship in case the Skipper wishes to check a point. A cynic once said that the most useless things aboard ship were lawn mowers and books on navigation and seamanship. The same might be said of stability books and fishing gear if you have not at least acquired some familiarity with them before you cast off.

It will be more than a little gratifying if lecturers at nautical colleges find this small volume a useful supplement or aid to their courses.

It is suggested that candidates for Second Hand (Special, Limited and Full) should be thoroughly familiar with all of Chapters 1, 2 and Appendix 7. In addition, they should be capable of answering simple questions on Chapters 4 and 10 and the effects of ice formation dealt with in Chapter 9. Candidates for Skipper (Full and Limited) require a sound knowledge of all the foregoing with the addition of a good working knowledge of Chapters 3, 5, 6 and most of the Dynamical Stability section of Chapter 8. The remainder is recommended reading, particularly the effects of damage dealt with in Chapter 9. Some matters of general interest and proofs of formulae, if required, will be found in the Appendices.

J. Anthony Hind

First Principles

Introduction

MORE considerations are involved in the safety of a ship than its stability, but all authorities concerned with safety of ships at sea are fully agreed that 'satisfactory stability' is the most important. As to what exactly constitutes this elusive virtue, i.e. the *standard* which should be laid down, is a matter of judgment and experience in any particular case. Inevitably, this means that there is room for argument on matters of important detail. Enthusiasm for safety and the laying down of standards of stability in both normal and damaged conditions has never been lacking, but *minimum* standards have only recently been adopted in official legislation. Paradoxically, this is fortunate because it would be more of a liability than an asset to end up with inflexible rules which might be applied without regard to the merits of any particular case.

The practical application of the *principles* of stability to ships is now a familiar operation and the factors which contribute to 'satisfactory stability' are universally acknowledged. These may be listed as watertight integrity, the immovability and proper distribution of weights (e.g. cargo) adequate freeboard, suitable metacentric height and dynamic stability, harmonious proportions of hull form, the ability to shed water from decks rapidly and, not least, the professional competence of the master and crew.

The student of stability may be encouraged to find that the fundamental principles are not too difficult to understand—but a newly acquired working knowledge should be accepted on the basis that where full and completely accurate studies are required, the subject is far from simple, especially in the investigation of

damaged conditions. However, for practical application in the
normal operation of ships at sea, the following exposition should be
adequate.

Tonnage

Every merchant ship has a number of different 'tonnages' and
it is important to know that some of these bear no relation to
weight at all. Gross and net registered tonnages are measures
of capacity (volume) in which 100 cubic feet has been artificially
selected to equal what is called 1 ton (i.e. 1 ton = 100 cu. ft. of
capacity). The figure of 100 makes multiplication and division very
easy in converting from gross or net tons to cubic feet, or vice-
versa.

Deadweight is measured in tons of 2,240 lb. weight (what
Continental countries call the English long ton). A ship's deadweight (dwt.) is the actual amount of weight in tons that the ship
can carry when loaded to the maximum draft permitted by
international law (but fishing vessels do not normally have to
comply with the International Loadline Convention, although
this is likely to be reviewed in the future). Deadweight includes
not only the cargo but also all movable stores such as provisions,
fuel, fresh water and feed water. On fishing vessels it frequently
also includes all movable fishing gear and tackle as well as the
crew and all their effects. So that:

Deadweight = Cargo (e.g. fish) + 'Consumables'

What the ship weighs completely empty is called the Lightship
or Lightship Weight. Therefore:

Lightship + Deadweight = Load Displacement

Displacement—The Principle of Archimedes

This Greek philosopher formulated the important principle
that a body wholly or partially immersed in a fluid loses weight
equal in amount to the weight of the fluid it displaces. This
applies whether the body is heavier, lighter or equal in weight to
an equal volume of fluid in which it is immersed.

It is fairly obvious that for a ship to float freely in water, the
weight of the ship must equal the weight of the amount (volume)
of water it displaces. The displacement can be expressed either in

tons weight or volumetrically (i.e. cu. ft.).* As 35 cu. ft. of seawater approximately equals 1 ton weight then:

$$\text{Displacement in seawater (tons)} = \frac{\text{Vol. of Displacement (cu. ft.)}}{35}$$

If the displacement in fresh water is required, then the divisor is 36. This, of course, is because the actual weight of any given volume is affected by its density.

$$\text{Seawater} \quad - \quad 1025 \text{ kg/m}^3$$
$$\text{Freshwater} \quad - \quad 1000 \text{ kg/m}^3$$

For this reason, a vessel's draft alters in going from a fresh water river to sea or vice-versa. The volume of displacement alters but not the weight. In other words, for any given displacement weight the drafts in salt water and fresh water are not the same. But in reading off a displacement against a draft from a displacement curve one must be sure as to whether the curve is for salt or fresh water—the weights are not the same. Sometimes the builder's hydrostatic particulars give only one curve. If so, it refers to seawater.

The change in draft due to change in density is obtained from the following formula:

$$\text{Draft change in inches} = \frac{\Delta (\delta_1 - \delta_2)}{\delta_2 \times TPI}$$

where, Δ = displacement (tons)

δ_1 = density (1) (oz./cu. ft.)

δ_2 = density (2) (oz./cu. ft.)

TPI is the tons per inch immersion (in metric, *TPCm*: tonnes per centimetre). This can be found from the hydrostatic curves supplied by the builders.

Tons per inch immersion is the weight required to cause a bodily sinkage of the ship of 1 inch. Suppose the area of the waterplane is A; 1 inch is 1/12 ft.

Then A (sq. ft.) $\times$ 1/12 (ft.) is the increase in underwater volume. This divided by 35 is the increase in displacement in tons.

---

*Or in metric tons (tonnes) and cubic metres (m³).

In using the metric version of any formula it is essential to use *consistent* units —lengths in metres, areas in sq. metres and weights in tonnes or decimals thereof.

Therefore, $\dfrac{A}{12} \times \dfrac{1}{35}$ is the tons per inch immersion in seawater,

i.e. $TPI = \dfrac{A}{420}$ or $\left[\begin{array}{l} TPCm\,(\text{SW}) = \dfrac{A}{100} \times 1.025 \\[6pt] \text{where } A \text{ is in sq. metres} \end{array}\right]$

Example 1

A fishing vessel has drafts of 9 ft. 6 in. forward and 10 ft. 6 in. aft. Bunkers and stores are then taken aboard of 20 tons and 12 tons respectively. Calculate new mean draft. $TPI = 8$

Weights shipped $= 20 + 12 = 32$ tons

$$\text{draft increase} = \frac{w}{TPI} = \frac{32}{8} = 4 \text{ in.}$$

$$\text{Original mean draft} = \frac{9.5 + 10.5}{2} = 10 \text{ ft.}$$

$\therefore$ New mean draft $= 10$ ft. 4 in.

Example 2

A large freezer trawler arrives off her home port with an even keel draft of 20 ft. corresponding to a displacement of 3,634 tons (density of seawater 1,025 ozs./cu. ft.). The waterplane area at this draft is 8,820 sq. ft. The vessel is directed to enter an enclosed basin in which the water density is 1,010 ozs./cu. ft. and having a depth of 20 ft. 10 in. alongside. Calculate the amount of cargo which must be discharged from an amidships fishroom into a lighter in order that the vessel can enter the berth with a ground clearance under the keel of not less than 12 inches.

Calculate increase in draft due to change in density:

$$TPI = \frac{\text{Area of } WP}{420} = \frac{8820}{420} = 21$$

$$\text{Increase in draft} = \frac{\Delta\,(\delta_1 - \delta_2)}{\delta_2 \times TPI} = \frac{3634\,(1025 - 1010)}{1010 \times 21}$$

$$= 2.57 \text{ inches.}$$

New draft in basin if no cargo discharged $= 20'\ \ 2.57''$

Limiting draft for 12" under keel clearance $= \dfrac{19' \ 10.00''}{4.57''}$
Excess draft

$$
\begin{aligned}
\text{Amount of cargo to discharge} &= TPI \times 4.57 \\
&= 21 \times 4.57 \\
&= 95.97 \ \text{(say 96 tons)}
\end{aligned}
$$

The water displaced by a ship is also known as the buoyant force or simply, buoyancy. The forces of ship weight and buoyancy must be equal and opposite (i.e. $W = B$). If they were not, the ship would either sink or fly! This leads to a consideration of where these two forces of weight and buoyancy act in a ship and two important definitions follow.

Centre of Gravity

This is the point at which the whole weight of the ship acts. It is the point at which the total weight of a rigid body may be assumed to be concentrated. Weight is a force which acts vertically downwards through or from the centre of gravity.

The position of the centre of gravity G entirely depends upon the distribution of weight in the ship. The height of G (the vertical centre of gravity, VCG) above some reference point, usually the keel line, is affected by topweight (raised) and by ballast (lowered). The actual effects can easily be calculated if the original position of G is known. This position can be found by conducting an inclining test or experiment and this is what the designer does when the ship is originally built. The Inclining Experiment is discussed in Chapter 3.

Centre of Buoyancy

This is the centre of gravity of the displaced water (i.e. the

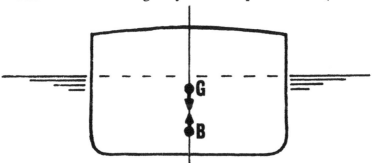

FIG. 1. Centres of Gravity and Buoyancy.

underwater volume of the ship) and is the point through which the resultant upthrust of the water surrounding the vessel may be considered to act. Fig. 1 illustrates the position diagrammatically. The position of *B* is entirely dependent upon the geometric form of the ship's underwater body.

Equilibrium

There are three states of equilibrium—stable, unstable and neutral.

A (homogeneous) cone standing on its base on a flat surface is said to be in stable equilibrium because if tilted slightly it would return of its own accord to its initial position.

A cone standing on its apex on a flat surface would be in unstable 'equilibrium' (not equilibrium at all in fact) because if disturbed it would continue to move further from its initial position. (We are not here concerned with spinning of the cone about its vertical axis whereby any alteration of the inclination of the axis of rotation is resisted by the gyrostatic moment.)

A cone on its side on a flat surface would remain in any position to which it was displaced and would therefore be in a state of neutral equilibrium.

It is the first two conditions with which we are intimately concerned. The third condition of neutral equilibrium has only limited application.

Initial or Metacentric Stability

INITIAL stability is the study or analysis of those conditions which determine the equilibrium of a floating body. It is usual to consider statical conditions, i.e. in the case of a ship, she is assumed to be freely floating upright in still water, and at rest.*

Fig. 2 shows the difference between a righting and an upsetting moment (a force multiplied by a distance is called a 'moment').

Consider a ship floating freely in still water and slightly inclined by some external and temporary force from the upright. Due to the change in shape of the underwater body, the centre of buoyancy B will move outwards from the centreline to a new position B_1. The position of the centre of gravity will not change from its centreline position (where great care has been taken by the ship designer to have it located). The two equal forces of weight and buoyancy ($W = B$), acting vertically and in opposite directions (and formerly acting along the centreline) will now be displaced horizontally by a distance GZ (called the Righting Lever or Righting Arm). A couple, $W \times GZ$ is thus formed which tends to rotate the vessel either back to its initial position (i.e. upright) or further from it in the direction of the original inclination. Equilibrium will not be regained until the couple has disappeared

---

*Stability calculations are made on the assumption of a waveless waterplane. At about maximum wavemaking speed in smooth water or in heavy head or following seas there can be the two situations of either wave crests at bow and stern with a trough amidships or a single wave crest amidships and hollows or troughs at bow and stern. The former case with the trough amidships is slightly beneficial from a stability point of view whereas the latter is unfavourable. The worst case is when a wave having a length the same as the ship is overtaking slowly. On average, wave action decreases righting moments (see also end of Chapter 5). A very adequate and detailed discussion of this subject for further study is given in a paper by J. R. Paulling, Jr. on 'Transverse Stability of Tuna Clippers', *Fishing Boats of the World* 2.

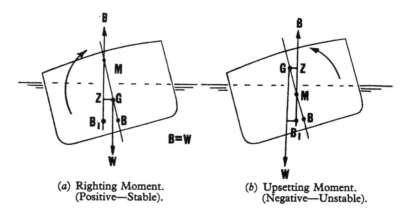

(*a*) Righting Moment.   (*b*) Upsetting Moment.
    (Positive—Stable).   (Negative—Unstable).

FIG. 2. Stable and Unstable Equilibrium.

and *B* and *G* are once more in the same vertical line. If the vessel does not return to the upright, but heels until *B* and *G* are in the same vertical line then the ship will heel to some permanent *Angle of Loll*. When heel is permanent it is called *List*.

It is important to note from an inspection of the diagrams that the direction of rotation of the vessel subsequent to initial inclination is dependent upon the relative positions of the centre of gravity *G* and the metacentre *M*. If *M* is above *G* the ship will be stable; if *M* is below *G* the ship will be *initially* unstable. If *G* is not in the centreline of the ship when at rest in the upright condition, or if *G* is a little above *M*, then the ship will loll. However, as *M* rises as the ship inclines, any further slight inclination will bring *M* above *G* and the ship will be stable again (see Fig. 3).

It may be stated, therefore, that *the metacentre is the limiting height to which the centre of gravity may be raised without producing initial instability.* Hence the term metacentre which means 'change point'.

Strictly speaking, the metacentre is defined by two co-ordinates so that we refer to the transverse metacentre and the longitudinal metacentre. It is the transverse metacentre with which we are mainly concerned here. The metacentre may be found at the intersection of a vertical line through the centre of buoyancy in the *initial position* with a vertical line through the centre of buoyancy in a slightly inclined position. This is the same as saying that a vertical line through the inclined centre of buoyancy intersects the median line of the body at the metacentre.

The distance *GM* is called the *initial metacentric height*, or simply metacentric height. Its amount is important, but it is no less important to remember that it is only in positions of equilibrium (i.e. at small angles of inclination from the upright, say at the most 7 degrees) that the relative heights of *G* and *M* are criteria of stability. For large inclinations the position of *M* will vary appreciably.

To summarise, the three conditions for stable equilibrium in still water are:

1. Buoyancy must equal Weight ($W = B$)

2. *B* must be in the same vertical line as *G*

3. *G* must be below *M*

$W \times GZ$ (tons ft.) is known as the *Moment of Statical Stability*.

It will be seen from Fig. 2 that $GZ = GM \sin \theta$, where θ is the angle of heel or inclination. From this we obtain the statement that the Moment of Statical Stability $= W \times GM \sin \theta$ (tons ft.).

The amount of metacentric height *GM* bears an important relationship to the period of roll of the ship and the acceleration of the motion. If *GM* is small there will be a condition of tenderness and the motion will be sluggish. From the stability point of view the vessel will be referred to as being 'crank'. The opposite of this is 'stiff' and excessive stiffness due to too much *GM* is not only extremely uncomfortable but could result in damage to fittings if not to the hull structure. Inadequate *GM* is extremely dangerous as it could result, in the worst case, in 'over-rolling' of the vessel in beam seas. The development of inadequate *GM* or its loss at sea due to negligence, damage or extreme conditions will be discussed later.

Negative Metacentric Height

A vessel with a negative *GM* is not necessarily in imminent danger of capsizing. In harbour or in calm waters the condition is not usually dangerous, but it is certainly a most undesirable condition and one which should be investigated without delay.

In such a condition the vessel will have a list. It cannot remain upright. The first sign that such a condition is being reached is a tendency for the vessel to 'flop' from side-to-side. With a few inches of negative *GM* the vessel will list to what has already been referred to as its Angle of Loll. At this point the vessel will have picked-up positive *GM*. In any further analysis this must now be considered as the *initial* position.

Fig. 3 shows an unstable vessel at its angle of loll. The centre of buoyancy is not on the ship's centreline. The centre of gravity will still be in the same position on the ship's centreline but the position of the metacentre will not be the same as for the upright condition. It is required to find the new position of *M*. By definition, *M* lies at the intersection of a vertical line through *B* (now the initial centre of buoyancy) and the vertical through the centre of buoyancy (B_1) in a slightly inclined position. It will be

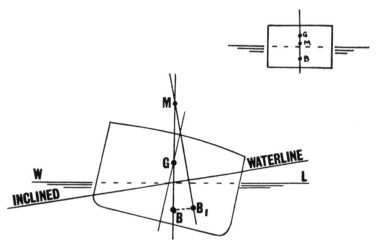

FIG. 3. Positive GM at Angle of Loll.
(*Small inset*: Vessel cannot remain upright in this condition.)

seen that the new metacentre *M* is above *G*, i.e. the vessel has positive metacentric height and a restoring couple will operate to return it to its angle of loll after any displacement or heel in the direction of the original list.

The causes of the list might be due to improper distribution of weight within the ship either too high or excessively on one side. There may be damage to the hull or slack water or a combination of any of these factors. The cause of list may be obvious, but if in doubt *assume instability*.

The following advice should be acted upon in any case of suspected instability:

1. Do not empty any fuel or water tanks below the waterline on the low side.

2. Press up all slack tanks to reduce liquid free surfaces as much as possible. This will most likely entail tank transfer.

3. Lower movable weights if possible, e.g. trim down fish in the hold and fishing gear.

4. Secure suspended weights and derricks and do not attempt to haul on the derricks.

5. As a last resort, ballasting (counterflooding) may be attempted. Put the vessel into the weather and 'heave to'. Maintain reasonable trim and start filling a tank about amidships. With a centrally divided double bottom tank it is most important to start filling on the *low* side first. This will, of course, make the list slightly worse for a short time but weight is being added as low as possible. When the tank is from about ⅓ to ½ full, carry on ballasting both sides together and press right up.

Effect of Weights on Ship Condition

The effect of weights can be considered under the following heads:

1. Raising or lowering weight already aboard.

2. Adding or removing weight.

3. Suspension of a weight (e.g. on a derrick)

4. Moving a weight transversely.

Weights Already Aboard

If a weight of w tons is raised a distance d feet the centre of gravity of the whole ship will also be raised to a new position G_1. The metacentric height GM will decrease by an amount GG_1.

Now the force w multiplied by the distance through which it is moved d is called the moment. This is equivalent to the weight or displacement of the ship multiplied by the effect (distance GG_1)

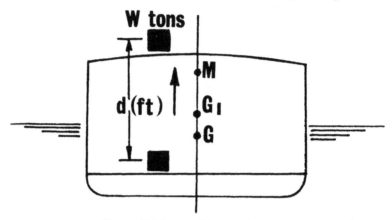

FIG. 4. Raising a weight already aboard.

on G. That is, $w \times d = W \times GG_1$.

$\therefore$ Loss of metacentric height $GG_1 = \dfrac{w \times d}{W}$

Where W (often designated by Δ) is the displacement of the ship in tons. Note there is no change in either draft or trim when a weight already aboard is raised. Lowering a weight has the reverse effect by increasing the GM.

Weights Added or Removed

Weights added or removed from a ship will cause:

a) Increase or decrease of displacement.

b) Increase or decrease of draft.

c) Alteration in trim fore-and-aft unless added amidships.

d) Raising or lowering of G.

e) Alter the position of B and M.

f) List, unless added on the centreline, but this is not significant for weights of moderate amount.

Although the positions of M and G will change there may, in fact, be no change in their relative positions in which case GM value would remain the same. An example will make the effect of adding and removing weights clear.

Example 3

On departure from the fishing grounds for home a trawler is found to have 30 tons of fish in the hold at a height of 8 ft. above the keel. Since leaving port the trawler has consumed 13 tons of fuel (c.g. 3 ft. above keel), 8 tons of fresh water (c.g. 6 ft. above keel), 2 tons of stores (c.g. 10 ft. above keel) and lost 3 tons of nets and fishing gear (c.g. 14 ft. above keel). Before leaving port the trawler had a displacement of 500 tons and the height of its centre of gravity above the keel *KG* was 7 ft. Calculate the new position of *G* above the keel. If in this condition the height of *M*, i.e. *KM* is 9 ft., what is the Metacentric height on leaving the fishing grounds and the value of the righting lever at 10 degrees inclination?

Item	Weight tons		Lever ft.	Moment tons ft.	
	+	—		+	—
Trawler	500		7	3,500	
Fish	30		8	240	
Fuel		13	3		39
FW		8	6		48
Stores		2	10		20
Nets etc.		3	14		42
	530 26	26		3,740 149	149
	$\Delta = 504$			3,591	

$\therefore$ (New lever) $KG = \dfrac{\text{Moment of } \Delta}{\Delta} = \dfrac{3,591}{504} = 7.14$ ft.

But $GM = KM - KG$

$= 9 - 7.14$

$= 1.86$ ft. on departure from fishing grounds.

Also, Righting lever $GZ = GM \sin \theta$

$= 1.86 \sin 10°$

$= 1.86 \times 0.1736$

$= 0.323$ ft. approx.

i.e. at 10° inclination the righting lever will be nearly 4 inches.

(Note: the sine of the angle of inclination is found from Trigonometrical Tables).

Suspended Weights

It is most important to realise that the centre of gravity of any freely suspended weight aboard ship does *not* act at the actual centre of gravity of the weight itself, but at the point of suspension. The point of suspension is therefore known as the *virtual* centre of gravity of the weight. A typically important example aboard a fishing vessel is the suspension of the loaded trawl from a derrick. So long as the trawl is off the deck its weight acts at the derrick head. This has a detrimental affect on metacentric height. Also if the point of suspension, the derrick head, is offset from the centreline there will be a heeling moment acting upon the ship. Fig. 5 explains the situation which is unaffected by any initial list of the ship.

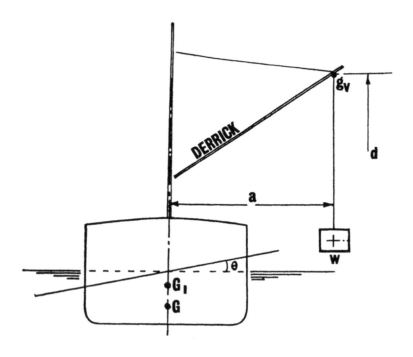

FIG. 5. Suspended Weight.

Metacentric height is decreased by GG_1

$$GG_1 = \frac{w \times a}{\Delta}$$

Angle of heel: $\tan \theta = \dfrac{w \times a}{\Delta \times GM}$, i.e. $\theta = \tan^{-1} \dfrac{w\,a}{\Delta\,GM}$

(Note: $\tan^{-1}$ means the angle whose tangent is . . .).

Angle of Heel Due to Moving a Weight Transversely

It is assumed here that the weight to be moved is already aboard the ship. If not, and the weight is to be added or removed from one side of the ship, then the problem is in two parts, viz. the weight is first considered to be added at the centreline (see previous relevant section) and then moved transversely—or vice-versa if being removed from the ship.

Consider a weight w to be moved a distance d transversely across the deck as in Fig. 6.

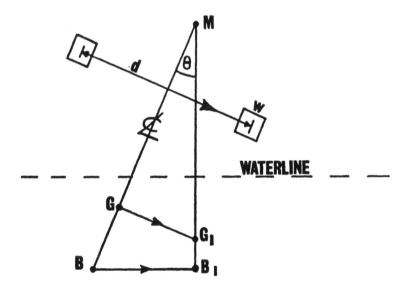

FIG. 6. Shifting a Weight Transversely (i.e. Angle of Heel due to Moving a Weight across a deck).

The transfer moment is $w \times d$

G will move in the direction of transfer to G_1

B will move similarly to B_1

The following relationships will hold:

$$w \times d = \Delta \times GG_1$$

$$\tan \theta = \frac{GG_1}{GM}, \text{ or } GG_1 = GM \tan \theta$$

$$\therefore \quad \tan \theta = \frac{w \times d}{\Delta\, GM}$$

$$\text{and} \quad GM = \frac{w \times d}{\Delta \tan \theta}$$

This last formula is important because it means that if the angle of heel can be measured as a result of moving a known weight a given distance then GM can be found. This formula is therefore made use of in the stability investigation known as an Inclining Experiment.

The Inclining
Experiment

IT is the practice of shipbuilders to carry out a stability investigation on all new ships (except those which are repeats of designs previously built). The object of this inclining experiment, as it is called, is to determine the position of the centre of gravity and the reason for a practical investigation is because the position of G is not calculable with any degree of reliability from design data alone.

The position of G in which the shipbuilder is interested is that for the Lightship Condition 'as built'. This is the fundamental position of the centre of gravity from which all other positions in any condition of loading of the ship can be derived by calculation if the loading information is known. The shipbuilder makes meticulous arrangements and rather lengthy and laborious checks to ensure the accuracy of the test, but a full description of these elaborate details is not neccessary here.

From the point of view of the ship's officer, the main interest in an inclining test is the value of the metacentric height. Any ship's officer could check a vessel's stability by means of a rough inclining test with quite simple facilities provided certain elementary precautions are observed. The whole operation could be set up and completed in a matter of a few hours.

The weather must be moderately good for an inclining test with the vessel moored head to wind and under the lee of warehouses or dock walls. Beam winds must be avoided and any appreciable movement of the water surface or beam currents would cause inaccuracy. The ideal, of course, is a drydock in which the ship can lie afloat perfectly still. Mooring ropes must be slacked-off, the shore gangway removed and, in fact, anything which would prevent the vessel heeling freely.

The ship should be reasonably upright in the first place and any loose weights (derricks, boats, etc.) should be secured in their normal stowage positions. Boilers should be at working level. Any slack oil or water having appreciable surface area will materially affect the result of the test. There are two schools of thought about the best way of dealing with this problem of tanks. One is that, as it is very difficult to be sure that tanks are either completely empty or pressed really 'hard-up', then all tanks should be very slack (about $\frac{1}{2}$ full). If this is done, it is argued, there is no doubt about free surface effect. It exists; and can be accurately allowed for. However, for ships' officers, it is probably easier to press all tanks hard-up and if this cannot be done with all fuel tanks then the slack should be limited to one or two and allowed for. Accurate soundings must be taken. Bilges must be pumped dry. Men aboard should be limited to those taking part in the test and they should return to the same position aboard after every shift of inclining weight until the pendulum swing has been noted.

Inclining weights can be of any suitable material, but pig-iron weights, which can be lifted and stacked by hand, are probably the best. The amount of weight required is that sufficient to incline the vessel about 1 to $2\frac{1}{2}$ degrees. More than 3 or 4 degrees inclination would be unsuitable because of the requirement that the position of M should not change during the test. To estimate the amount of inclining weight required, take the displacement in tons and divide it by the half-beam of the ship in ft. The answer is the inclining weight in cwt. These should be placed in two exactly equal piles one on each side of the ship as near midships as possible. Put a chalk mark round their positions. Measure their distance apart centre-to-centre. This is the shift of weight d.

One plumb line is sufficient, but shipbuilders use two to average out the errors and as an additional check. It should be as long as possible and protected from the wind. It is best to suspend the line above and through a hatch with the bob immersed in a wide bucket or trough of water to provide damping. A horizontal wood batten is fixed just above the bucket so that the plumb line very lightly rubs its edge. The length of the plumb is accurately measured from point of suspension to the top edge of the batten.

Take the drafts fore-and-aft. Mean draft will give the ship's displacement in tons from an inspection of the vessel's displacement curve supplied with the builder's particulars. Note the density of the water in which the ship is floating by hydrometer. It is necessary to get a sample of the water from about amidships in a bucket in order to be able to read the hydrometer properly.

Mark the centreline of the plumb line on the wood batten with a pencil. Proceed as follows (see Fig. 7).

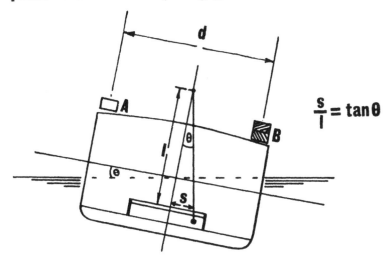

$$\frac{s}{l} = \tan\theta$$

FIG. 7. Inclining Experiment.

1. Move weight *A* to position *B*. Note deflection of pendulum and mark on batten.
2. Move weight *A* back to original position. Pendulum should return to centreline position on batten or nearly so.
3. Move weight *B* to position *A*. Note deflection.
4. Return weight *B* to original position.
5. Repeat moves 1 to 4.

In theory, each of the deflections of the pendulum to port and starboard should be identical. In practice this will not be so, which is the reason for taking several readings and averaging. Also, the plumb line will not come to rest completely so that it is necessary to estimate the centre of its swing when marking the deflection on the batten. The plumb bob must not touch the side of the bucket when a reading is being taken. An assistant should be on hand to move the bucket if necessary as the ship inclines.

The following data is now available:

Δ = Displacement of ship in tons (from mean draft)

w = Inclining weight (half the total) in tons

d = Shift of inclining weight in feet

l = Length of pendulum (plumb line) in inches

s = Deflection (average) of pendulum across batten in inches

δ = Density of dock water. (If the density varies appreciably from that of seawater at 1,024 oz./cu. ft. then it may be necessary to correct the Δ for density, see Chapter 1.)

Now $\dfrac{s}{l}$ = $\tan\theta$, where θ is angle of inclination and

$$GM = \frac{w \times d}{\Delta \tan \theta} \text{ (see end of Chapter 2)}$$

If the height of the centre of gravity (*VCG*) of the ship in the inclined condition is required, it is necessary first to find the height of the metacentre from the builder's particulars at the draft or displacement obtaining. Then, $KG = KM - GM$. (Note that the builder's *KM* will be for even keel conditions.)

The particulars thus found include, of course the effect of the inclining weights. This makes no appreciable difference but if it is wished to correct for Δ, *VCG*, *GM* etc., the method is as described in Chapter 2 under the heading "Weights Added or Removed".

Inertia and Free Surfaces

Waterplane Inertia

THE subject of ship stability involves consideration of the moment of inertia of the ship's waterplane and of the surfaces of any liquids within the ship or ship's tanks which are free to move. Inertia is the resistance of a body to change its state of rest or uniform velocity in a straight line (Newton's First Law of Motion). When we are dealing with rotations about some axis, as we are in the case of ship inclinations (angular motions), then it is the *moment of inertia* which has to be measured.

With transverse inclinations we are usually concerned with the moment of inertia of the ship's waterplane about its longitudinal axis, i.e. the centreline. The moment of inertia is found by measuring the half ordinates of the waterplane and putting their cubes (i.e. $\frac{1}{2}$ ordinate³) through Simpson's multipliers. This gives products (or moment functions). These products are then summed. This sum then has to be multiplied by two for both sides of the *WP* and by the common interval between the ordinates as well as by fractions of one third and either one third or three eighths depending upon which of Simpson's Rules has been used. The first fraction of $\frac{1}{3}$ derives from basic principles of waterplane inertia which involve the methods of the integral calculus. A theoretical treatment of the subject for the more advanced reader will be found in Appendix 2.

In general, moment of inertia about the ship's centreline may be stated thus:

$$I = 2 \times \tfrac{1}{3} \times \tfrac{1}{3} \times \text{Common Interval} \times \text{Products for } I$$
$$= 2/9 \times CI \times P \text{ (Using Simpson's First Rule)}$$

(*N.B.* If measurements in feet the results are in ft⁴ units.)

Metacentric Radius

The position of the metacentre M depends entirely on the shape and dimensions of the ship's waterplane and the underwater form of the ship. The formula:

$$BM = \frac{I}{V}$$

is one of the most important in ship stability. It states that *the separation of the transverse metacentre and the centre of buoyancy equals the moment of inertia of the intact waterplane about the centreline divided by the underwater volume of the ship.*

From this it follows that so long as the draft remains constant, the height of the metacentre (above any reference point) depends mainly on the beam of the ship or the *intact* waterplane. Loss of waterplane inertia therefore has a serious effect on a ship's stability.

Shallow draft in relation to beam gives high metacentres and therefore large initial stability. This rapidly disappears at large inclinations where a deck edge goes under and the bilge comes out of the water. For this reason, rafts can be dangerous for the unwary. A derivation of the above formula is to be found in Appendix 3.

BM is known as the *Metacentric Radius*. It is related to initial stability by the following equations:

$$GM = KM - KG$$
$$\text{or } GM = KB + BM - KG$$

Example 4

A rectangular pontoon 60 ft. long with a beam of 20 ft. has a displacement of 300 tons with its centre of gravity 8 ft. 6 in. above the keel. Calculate the initial stability condition in seawater and the final *GM* after the addition of 50 tons of cargo 2 ft. above the bottom of the pontoon.

Volume of displacement $= 35 \times 300 = 20 \times 60 \times$ mean draft

$$\therefore \text{draft} = \frac{10500}{1200} = 8.75 \text{ ft.}$$

$$BM = \frac{I}{V} = \frac{60 \times 20^3}{12 \times 35 \times 300} = \frac{80}{21} = 3.85 \text{ ft.}$$

For a box form KB equals half the draft, i.e. 4.375 ft.

Then, $KM = BM + KB = 3.85 + 4.375 = 8.225$ ft.

$$GM = KM - KG = 8.225 - 8.5 = -0.275 \text{ ft.}$$

The pontoon has a negative metacentric height and is unstable.

Add 50 tons cargo:

Weight	VCG	Moment
300 t	8.5 ft.	2550
50 t	2.0 ft.	100
350 t		2650

New $KG = \dfrac{\text{Moment of weight}}{\text{Weight}} = \dfrac{2650}{350} = 7.57$ ft.

$$\text{New } GM = \text{(old) } KM - KG = 8.225 - 7.57$$
$$= 0.655 \text{ ft.}$$

The Effect of Liquids having a Free Surface

Slack water or oil in the tanks of a ship will remain level when the ship heels, i.e. the surface of the liquid in a partially filled tank will remain parallel to the waterline. This has an adverse effect on the stability of the vessel.

Where a tank is completely filled no movement of the liquid is possible and it behaves just as a fixed solid with its centre of gravity at its centre of volume.

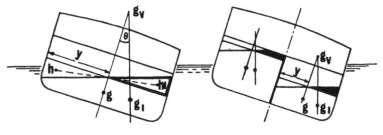

(a) Deep Tank.　　　　(b) Deep Tank centrally divided and filled to different levels.

FIG. 8. Free Surface Effects of Contained Liquids.

Fig. 8a shows a deep tank partially filled with water ballast. When the vessel is upright the centre of gravity of the ballast is at g. When the vessel is inclined the centre of gravity moves to g_1, i.e. the weight of the ballast now acts vertically downwards through g_1 and this line of action cuts the original line through g at a point g_v known as the *Virtual Centre of Gravity* of the liquid. The effect of the free surface therefore is the same as if the mass of liquid had been transferred from g_1 to g_v (or g to g_v). The transfer moment is therefore $w \times gg_v$. The effect on GM is similar to the problem of suspended weights dealt with in Chapter 2 so that,

$$\text{Loss of } GM = \frac{w \times gg_v}{\Delta}$$

Unfortunately it is not readily convenient to determine the distance gg_v so the transfer moment of the water as the ship heels is considered instead.

It can be shown that the wedge volume of water transferred is $\frac{2}{3}\Sigma y^3 \delta x \times \delta\theta$ where y is half the width of the tank and $\delta\theta$ is the angle of inclination. Now $\frac{2}{3}\Sigma y^3 dx$ is the moment of inertia of the (free) surface i about its fore-and-aft centreline, i.e.

$$\text{Wedge volume transfer moment} = i \times \delta\theta$$

but this transfer moment clearly also equals $v \times gg_1$, where v is the total volume of all the ballast in the tank.

$$\therefore v \times gg_1 = i \times \delta\theta$$

but $\quad gg_1 = gg_v \times \delta\theta \quad$ (since $\delta\theta$ is small sin $\theta \approx \delta\theta$)
substituting $v \times gg_v \times \delta\theta = i \times \delta\theta$

$$\therefore gg_v = \frac{i}{v}$$

as above, Loss of $GM = \dfrac{w \times gg_v}{\Delta}$ or $\left[\dfrac{v \times gg_v}{V}\right]$

substituting, Loss of $GM = \dfrac{v \times \dfrac{i}{v}}{V}$

$$= \frac{i}{V}$$

where $V =$ underwater volume of ship.

If the density of the liquid in the tank (e.g. oil fuel) is different from that of the water in which the ship is floating, then this result must be modified in the ratio of the two densities, then

$$\text{Loss of } GM = \frac{i}{V} \times \frac{\delta_t \text{(tank)}}{\delta_s \text{(sea)}}$$

This is known as the correction for free surface and this simple formula is of vital importance. It applies irrespective of the depth, size or position of the tank or its shape and it applies to any loose water anywhere in the ship. The free surface effect depends upon the surface area of the liquid so that a few inches of depth will have the same effect on initial *GM* as a large volume. Where liquid is taken aboard whereby the displacement increases then the added weight (considered as a solid) effect on *GM* must be considered separately. The effect of several slack tanks is cumulative, i.e. the effects are summed in applying the total free surface correction.

At fairly large angles of inclination, the free surface effect is diminished for small depths or large depths of liquid, i.e. it depends on the amount of liquid in the tank (see Fig. 8b).

In general, the greatest free surface effect will be mid-way, i.e. tank half-full condition. It is also important to bear in mind that as surface inertia is expressed in ft^4 units then dividing the width of the surface area into two equal parts gives an inertia of each part only $\frac{1}{8}$ of the undivided surface or a total for *both* parts of $\frac{1}{4}$ of the undivided surface.

(Taking the general formula $I = \dfrac{LB^3}{12}$

$\dfrac{l(b/2)^3}{12}$ is only $\frac{1}{8}$ of $\dfrac{lb^3}{12}$ where b is the width of the surface).

This means that in ballasting a double bottom tank with a watertight central division, the free surface loss of *GM* is only $\frac{1}{4}$ what it would have been with the tank undivided (see Fig. 8b).

Example 5

A trawler displacing 500 tons in seawater has a *GM* of 2 ft. and a draft of 8 ft. Fuel oil having a density of 0.9 is being drawn from a double bottom tank 30 ft. long, 24 ft. wide and 3 ft. deep. Calculate approximately:

a) The *GM* when the tank is half-empty.

b) The *GM* for the same condition but assuming that the tank has an oiltight centreline division.

(Take density of seawater at 1,025 ozs./cu. ft.)

$$\text{Oil used} = \frac{30 \times 24 \times 1.5 \times 900}{2240 \times 16} = 27 \text{ tons approx.}$$

(The loss of weight from the D.B. tank is equivalent to a gain in buoyancy which would cause a bodily rise of the ship, i.e. a layer of buoyancy at the *LWL*. The distance of the c.g. of the lost weight from the *LWL* is therefore the approximate lever of the moment of this lost weight. Similar arguments apply where a D.B. tank is being filled or ballasted.)

Distance of c.g. from *LWL* of the fuel used from D.B. tank:

$$8 - 2.25 = 5.75 \text{ ft. (d)}$$

$$\text{Loss of } GM \text{ (solid)} = \frac{w \times d}{\Delta} \quad (\Delta = 500 - 27 = 473 \text{ tons})$$

$$= \frac{27 \times 5.75}{473}$$

$$= 0.33 \text{ ft. nearly}$$

Loss of *GM* due to free surface of oil in tank

$$= \frac{i}{V} \times \frac{\delta t}{\delta s}$$

$$= \frac{30 \times 24^3}{12 \times 473 \times 35} \times \frac{0.9}{1.025}$$

$$= 1.83 \text{ ft. approx.}$$

Final $GM = 2 - (0.33 + 1.83) = -0.16$ ft.

a) When the tank is half-empty, there is a negative *GM* of 0.16 ft. which is most unsatisfactory.

b) If the D.B. tank is subdivided centrally the final *GM* would be: $2 - \left(0.33 + \dfrac{1.83}{4}\right) = 1.21$ ft. (positive).

Stability at Large Angles

INITIAL stability of ships has been shown in Chapter 2 to depend on the metacentric height *GM*. At angles of heel greater than about 10°–15° the value of *GM* alone is no longer an adequate criterion of stability. This is because the metacentre *M* where the upthrust of the buoyancy in the inclined position B_1 (Fig. 2) intersects the middle line is no longer fixed. It is not possible to determine the point *M* by simple formula and it is necessary to make rather tedious calculations to determine the stability levers (righting levers) *GZ*. The product ΔGZ is known as the righting moment.

However, if at any given displacement (condition of loading) the ship is successively inclined and the *GZ* levers found for

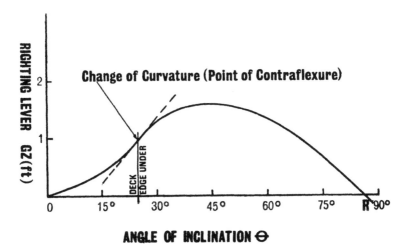

FIG. 9. Statical Stability Curve.

each inclination θ then these values can be plotted in the form of a curve on a base of angles of inclination. This is known as a *Statical Stability Curve* and it has the general form shown in Fig. 9.

Attwood's Formula

As stated above, the righting moment or moment of statical stability is Δ*GZ*. Now referring to Fig. 10 we see that

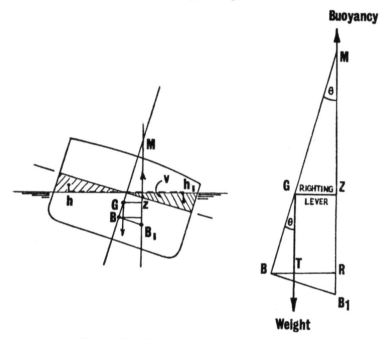

Fig. 10. Stability at Large Angles of Inclination.

GZ = *BR* — *BT* (where *BR* is horizontal transfer of the centre of buoyancy). And the moment of transfer of the wedges is:

$$v \times hh_1 = BR \times \triangledown$$

where v = volume of wedge; hh_1 = horizontal transfer of wedge and $\triangledown$ = volume of ship's displacement.

$$\therefore BR = \frac{v \times hh_1}{\triangledown}$$

Also $BT = BG \sin \theta$

$\therefore GZ = \dfrac{v \times hh_1}{\nabla} - BG \sin \theta$

$\therefore$ Moment of Statical Stability $= \Delta \left[\dfrac{v \times hh_1}{\nabla} - BG \sin \theta \right]$

This is known as Attwood's Formula. Due to the form of the ship the immersed and emerged wedges are not equal at large angles of inclination so some correction must be made to Attwood's formula. This is usually considered as a layer correction either added to or subtracted from the displacement. It has also been assumed that transverse inclination does not cause a change of trim. Actually this is not so, due to variation in wedge volumes from fore to aft so there is slight change in trim. However, such precision is not required for practical purposes and the work involved in calculating wedge moments of transfer is extremely laborious from the point of view of the ship's officer. The foregoing background explanation has been necessary in order to explain the source of derivation of *Cross Curves of Stability* the use of which is extremely simple and of great value in obtaining a picture of a ship's stability condition fairly quickly.

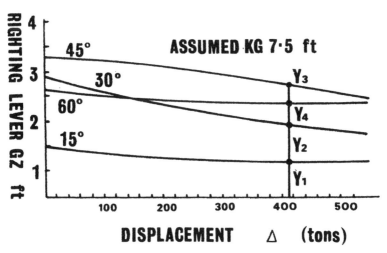

FIG. 11. Cross Curves of Stability.

Cross Curves of Stability

GZ levers or righting moments Δ*GZ* are obtained from Attwood's formula or some modification thereof. As explained at the beginning of this Chapter, these can be used to draw a statical stability curve for any given displacement. What is done in practice, however, is first to draw cross curves of stability which are curves of *GZ* for *constant* angles of heel on a base of varying displacement. The shipbuilder normally supplies a set of these cross curves for a range of displacement from the light to the full load condition. This enables a statical stability curve to be obtained very simply and quickly by graphical means for any displacement, i.e. condition of loading, that needs to be investigated. Fig. 11 shows a set of cross curves for a range of angles of heel plotted on a base displacement.

If there is a significant departure in any actual operating trim from that assumed when preparing the usual standard cross curves, then the shipbuilder's design department is now required under British Rules to repeat the cross curves data for the widely different trim conditions. This is because fixed centres of buoyancy and flotation and fixed values for longitudinal metacentre, tons per inch (or tonnes per centimetre), and moment to change trim at any given displacement apply only for moderate trim changes. There is also the added complication, mentioned in the previous section, that a ship cannot heel without trimming slightly although this is of more interest in respect of the ship's motion than for the effect it has upon transverse stability.

At this point the actual calculations for rigorous stability studies become complex and tedious and scarcely practicable without the facilities of a computer; nor are they of any interest to ship's officers, though instructions based upon their conclusions could be. However, for those who have to perform the calculations, and the more advanced student, an understanding of some of the interrelationships is desirable.

Suitable computer programs (not 'programmes' which are something else) are now available which enable stability data to be calculated on what is sometimes referred to as the 'free' trim basis. This method will henceforward be applied to all new vessels built under British Rules.

For large or excessive trim conditions where the usual hydrostatic curves do not afford reliable data it has always been possible to approach the situation from the other end, i.e. given the trim, what are the derived hydrostatics? These can then be related to calculated centres of gravity to output stability data. Inclined waterplanes or, indeed, any curved or irregular waterplane can be set-off on a series of Bonjean Curves (a description of which will be found in any elementary text book on naval architecture) and integrated for displacement, moments and centres in the usual way. These will give precise information for the condition considered and not usually obtainable direct from standard hydrostatic curves —and the method is invaluable not only for large operating trims at sea but also in damaged stability investigations (flooding), launching and structural strength studies.

The whole of this book is devoted to the rotation of a vessel about the two horizontal axes, i.e. that about the centreline longitudinal axis of the waterplane which is transverse (heeling and rolling) and that about a transverse axis through the centre of flotation of the waterplane (trimming or pitching). It is usual to treat the two as independent of each other. Whilst this is true of longitudinal inclinations, it is not strictly true of the transverse inclinations as applied to all normal seagoing ships which have unsymmetrical waterplanes about amidships fuller aft than forward. As the vessel heels or rolls the centre of flotation moves aft, because the aft unsymmetrical fullness increases in relation to that forward, and the centre of flotation is no longer on the fore and aft centreline of the original waterplane. This induces trim or pitching to maintain equilibrium. However, this would not happen with a double-ended vessel which is symmetrical about amidships (like some barges, lighters and pontoons). On the other hand, all vessels are symmetrical about their fore and aft centrelines at any given waterplane (draft). The centre of buoyancy must remain, therefore, on the middle line when the ship trims or pitches; in which position it cannot produce a heeling moment. This is why a vessel can pitch in a head sea without rolling but cannot roll in a beam sea without some pitching. Moreover, a vessel cannot do either without heaving (q.v. page 94) because of the inequality of immersed and emerged wedge volumes and the necessity to maintain constant displacement which involves changes in draft and, therefore, a vertical oscillation.

Statical Stability Curves

In order to construct a curve of statical stability draw a vertical line through the cross curves from the displacement in question. Measure or 'lift' the ordinates Y_1, Y_2, Y_3 etc. from the base line. These are the GZ values at the angles of inclination. Set these distances up on a base of angles of inclination and draw in the statical curve proceeding as in Fig. 12.

   a. Draw RH part of Curve and read-off GZ value at $57.3°$.

   b. Calculate approx GM, $GM = \dfrac{GZ}{\sin 57.3}$

   c. Set-off GM on perpendicular and draw in the tangent to the
      curve at the origin O.

   d. Complete the curve.

Obviously the GZ levers at small angles of heel depend mainly on the metacentric height. The tangent to the statical curve at the origin O can be drawn by erecting an ordinate at $57.3°$, measuring off the metacentric height and connecting this point to O.

One note of warning must be sounded before proceeding to draw a statical curve with ordinates obtained from cross curves. Cross curves are constructed for an assumed position of centre of gravity KG. Therefore any ordinate is only correct for this assumed position of G. The assumed KG will be noted on the cross curves and if the actual KG differs from this, the 'lifted' ordinates must be corrected by $GG_1 \sin \theta$. In other words, the difference in actual and assumed KGs multiplied by the sine of each angle of inclination in turn must be either added to or subtracted from the GZ ordinates. This is very easily applied.

Statical stability curves, so easily derived from a set of cross curves, are most descriptive diagrams of the stability characteristics of the ship in any condition of loading. They show (Figs. 9, 12 and 14):

   1. The value of max. GZ and the angle at which this occurs.

   2. The range of stability OR. R is known as the vanishing
      angle.

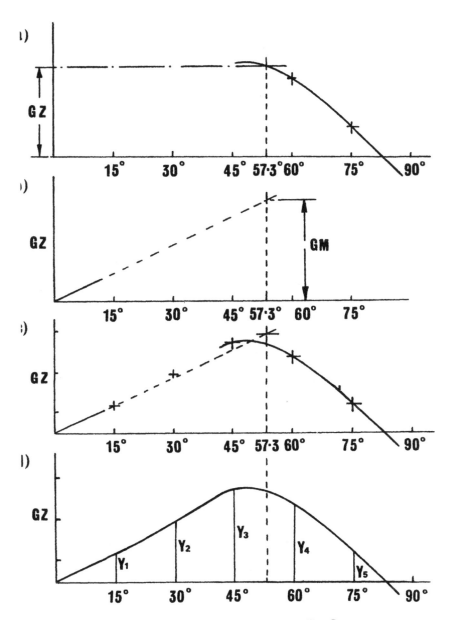

FIG. 12. Plotting and drawing a Statical Stability Curve.

3. The angle at which the deck edge goes under.

4. The approximate *GM*.

5. The dynamical stability.

Further consideration of these points is of great interest. External forces at work upon the ship are due to the action of wind, waves, out-of-balance weights, etc. If these forces are applied steadily their inclining moment (upsetting moment) will balance the corresponding righting moment ΔGZ at some angle of heel *at which the ship will be held*. If the external forces are not steady then the ship will have a positive righting moment at any angle up to the vanishing angle at which ΔGZ is zero after which the ship will generally capsize. It is also important to note the angle of maximum *GZ* where the righting moment ΔGZ is also of course, a maximum. If a *steadily* applied upsetting moment exceeds this value *the ship is in danger of capsizing*.

There is no point in continuing stability curves beyond 90° when the ship is on her beam ends. In practice, where the vessel would sink due to flooding through openings, the statical curve should be cut off at the corresponding angle to indicate entire loss of stability.

The angle at which the deck edge goes under can be estimated from the statical curve at the point where the *rate of change* of *GZ* goes from positive to negative. In other words, although *GZ* is still increasing the rate of increase is beginning to slow down. This is the middle of the *S* of the curve (see Fig. 9). Angle of deck edge immersion can, however, be accurately obtained from a section or Body Plan of the ship by drawing the inclined water-line.

The dynamical stability or work done is measured by the area under the statical curve up to the angle considered.

FACTORS INFLUENCING THE SHAPE OF
STABILITY CURVES

Beam
It has already been noted in Chapter 4 that a ship's beam (in relation to draft) has a powerful effect upon the height of the

metacentre. If length and draft are constant, then I/V varies as the square of the beam measured at the waterline. The improvement in stability consequent upon increase in beam does not, unfortunately, fully extend to the larger inclinations. Fig. 13 shows the general result.

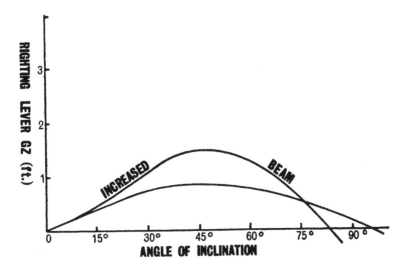

FIG. 13. Effect of Beam on Stability.

(i) GZ increased.
(ii) Range of Stability reduced. } No alteration in position of G.

Freeboard

Amount of freeboard is probably the most important factor affecting the stability of a ship. It is therefore of very considerable importance in the safety of a ship especially if that ship is vulnerable by reason of weather deck openings to entry of water to below decks.

It will be quite obvious that 'height of side' or freeboard has a considerable bearing on the angle at which the deck edge goes under (and at which waterplane moment of inertia is lost). At small angles the freeboard has no effect on stability, but from the point at which the deck edge goes under in the low freeboard vessel, the stability curve for the increased freeboard vessel will break away to increased righting levers. Not only will the maximum *GZ* be increased but it will also occur at a greater angle of heel. It will be evident that low freeboard associated with small maximum *GZ* and short range could be a source of danger. If for any reason nets and fishing gear have to be carried on the weather deck (say on the run home), it may be possible to stow them in such a way that the freeboard is virtually increased with beneficial effect on the range of stability.

It will be appreciated that increased freeboard has a more beneficial effect on stability than beam but a change in draft and

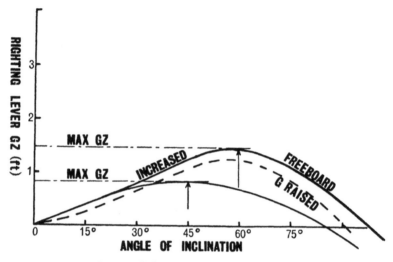

FIG. 14. Effect of Freeboard on Stability.
(i) Increased Righting Lever GZ. (ii) Max. GZ on greater Angle of Heel.

therefore displacement will normally be accompanied by a change in the height of *G*. (As also would any difference in depth of ship in comparing vessels.) Increased freeboard therefore would usually be associated with an increase in the height of *G*. The dotted curve in Fig. 14 shows the correction to the upper curve for the rise in *G*.

Height of G

A change in the height of *G* is shown in Fig. 15. We have already seen this influence in the last illustration.

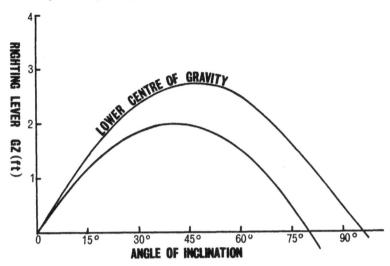

FIG. 15. Change in Height of Centre of Gravity.

The effect of lowering *G* is to improve stability at both large and small angles and increase the range. It will be appreciated therefore that a skipper has considerable latitude of choice in the matter of the stability of his vessel to the extent that the position of *G* is dependent upon how the vessel is stowed. No alteration in the dimensions of the vessel is possible but amount and distribution of cargo is to some extent a matter of choice. In this way the draft/freeboard relationship and the position of *G* in relation to *M* can be controlled.

Ballasting lowers *G* and improves stability but also increases displacement, deepens the draft and reduces freeboard. These latter may affect the stability unfavourably at large angles though the net effect is usually beneficial. Ballasting is often ineffective in deeply loaded vessels with considerable top hamper. The only remedy for this would be to remove the topweight.

STABILITY STANDARDS AND
MINIMUM STABILITY

As pointed out in the Introduction to Chapter 1, the *criteria** of ship stability and their practical application are now universally accepted. There is, however, less agreement as to what the values of the *parameters** should be for any particular class of vessel; though all are agreed that one set of parameters would not apply to all types and sizes. There are, however, numerous proposed parameters (several of which values are given throughout this book) which have merit and, in the absence of complete agreement, it is worthwhile making the best of them. There is no simple answer and no royal road to the understanding of what is, after all, rather a complicated subject. This is especially so when general advice has to be given rather than advice based upon a particular case where all the factors can be assessed and treated on their merits. The variables and the probabilities are too numerous for much specific advice and, indeed, this is the reason why all shipmasters need to have some basic knowledge of the factors involved and how to base their judgements upon them.

The question of *minimum* parameters for the various stability criteria in any particular operating condition for different types of vessel, and which are now the subject of legislation, is another matter. These are intended to be limits for safe operation not proposed optimum parameters. Whilst official minima are given later in this Chapter, it is first necessary to concentrate on the principles by which they can be understood.

We know a little about capsizing, but actual cases are difficult to reconstruct and there is hardly a mass of detailed evidence the reliability of which is beyond suspicion. But we do know that the moments righting the ship must be considered in relation to the heeling (upsetting) moments; and that so long as there is stable equilibrium between these two moments, the ship will not capsize. By implication, the levers are also known.

Metacentric height, *GM* is not, by itself, an adequate measure of stability. Minimum righting levers, or freeboard or dynamic stability or limiting heel or any other single simple factor by itself is quite inadequate as a measure. All these factors require to be taken into consideration in relation to all the others. Even

---

*A *criterion* is a principle or quality upon which a judgement can be based such as freeboard or the theory of dynamic stability. Of course, a criterion has then to be quantified so that a *parameter* is a measurable quantity, constant in the case considered, but which may vary in different cases; the two go together. Thus, whilst all are agreed that freeboard is an indispensable quality for ship stability, just how much is required in terms of distance may be a matter of opinion and/or law.

then, there is difficulty in deciding what the particular figures should be—it is a quantitative problem as well as a qualitative one. Nor is it simply a question of providing as much stability as possible, because that would probably be too much. So any general recommendations which can be given are necessarily tentative and probably approximate. It all depends upon individual cases and the best advice that can be given is that any advice must be treated with caution and applied with understanding.

In spite of all criticisms and attempts at simplification the most picturesque description of a vessel's stability still seems to be the fundamental cross curves of stability and statical stability curve. It cannot be over-emphasised that this must be set against the likely upsetting moments so that an appreciation of the stability balance can be obtained. This is the principle of the matter which must be understood by all seagoing officers. Without it, stability factors are not meaningful. It is substantially true to say that the problem of stability is the problem of the small ship. Even big trawlers are small ships relative to most ocean-going vessels and stability increases as the fourth power of the scale so absolute size is important and a big ship is inherently stiffer than a small one.

A ship is a solid three-dimensional body so if her length, breadth and draft are all doubled she is not of course double the previous size—she is eight times larger! The product of her linear factors is cubed, i.e. $L \times B \times d = \triangledown$, becomes $2L \times 2B \times 2d = 8\triangledown$. If this latter is then multiplied by the doubled righting lever (we are doubling linear dimensions) to give the new righting moment the result is a stability sixteen times what it was before (i.e. $\triangledown \times GZ$ becomes $8\triangledown \times 2GZ = 16\triangledown.GZ$). The converse is equally true. (In fundamental units of mass (M), time (T) and space (S or L) we have multiplied four linear units together, i.e. L^4).

The other side of the balance sheet is taken up by the moments tending to heel or upset the ship. Although contained water (the shipping of heavy 'green' seas) on the larger vessel would probably also vary as the fourth power of the scale, because of volume, lever and free surface effects, this is not the case with windage. This latter is basically related only to exposed projected area and a lever (q.v. page 77 et seq.) so it varies only as the cube of the scale. It might be mentioned in passing that herein lies one of the fundamental problems which limited the development of the conventional sailing ship. The space required on and above deck to put the vast and complicated sail area necessary to propel the much increased displacement of the slightly longer hull at reasonable sailing speeds could not finally be obtained. The discrepancy in the rates of increase of the two parameters was too much. We have all seen pictures of these unseaworthy rigs with their enormous too tall masts festooned with clouds of canvas to the extent of ridiculous skysails and stuns'ls which were the result of the economics of scale versus the power of the scale!

Freeboard

All merchant ships over 500 gross tons other than private yachts and fishing vessels are required to have a loadline fixed by international law. It has been universally agreed that there is a point beyond which a vessel may be considered to be overloaded; the freeboard is inadequate and the ship is therefore unsafe at sea.

Overloading is not uncommon in fishing vessels. Fishermen are inclined to disregard any lack of deadweight capacity if there is a heavy catch to be obtained. However, the reasons for this are beyond the scope of this book, though it might be observed that a fixed loadline would be as unpopular with fishermen as it would be difficult to enforce.

Nevertheless, a loadline (minimum freeboard) for fishing vessels agreed on an international basis is felt to be needed by experts in some countries. Some Administrations have already fixed what they consider to be safe loading limits. A common criterion is to fix the freeboard as some fraction of the depth of hull (e.g. D/10). A somewhat more scientific recommendation (and one which gives rather higher freeboards than the example quoted) is that, for flush-decked fishing vessels, the freeboard should not be less than that required at a deck edge immersion of $12\frac{1}{2}°$. In other words, the deck edge should not immerse at an angle of heel less than $12\frac{1}{2}°$. This is the same as saying $f = \frac{1}{2}B \tan 12.5°$, where B is the beam of the ship. The resulting freeboards are shown in Fig. 16. On the vexed question of legislating for freeboards or not, perhaps the most balanced view is that the skipper should thoroughly understand its general effect upon stability and its particular effect upon his own vessel in terms of the minimum

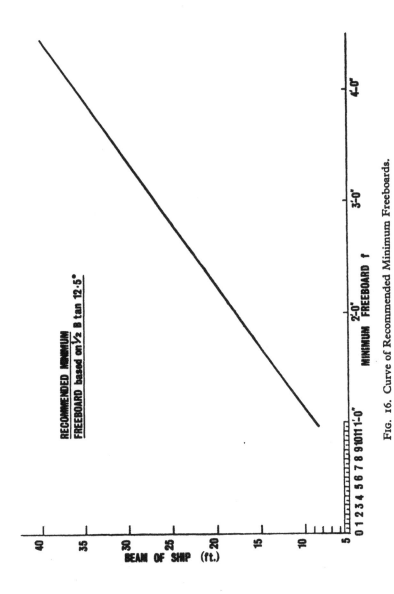

FIG. 16. Curve of Recommended Minimum Freeboards.

recommended freeboard and then leave the amount to his discretion in the light of all the circumstances at the time. A great deal depends upon his level of confidence in his own capabilities and experience, the weather conditions, the locality and the proximity of the vessel to port as well as the seagoing abilities of his vessel. Statutory rules, therefore, should be restricted to constructional features that can be incorporated during the building of the ship such as will restrict ingress of water and flooding like closing appliances and coaming heights and those that will rid her of loose water like freeing ports.

Most authorities seem to agree that the metacentric height of a fishing vessel in the light condition should be more than 1.25 ft. and not less than 2 ft. at deep load. The minimum *GM* for purse seiners has been recommended at 1.48 ft. Another figure given for trawlers in the worst stability condition is a *GM* of not less than 1.31 ft. For near water fishing vessels, general confidence has been expressed in righting levers (*GZ*'s) of about 12″ at 30° to 40° inclination and associated with metacentric heights of about 18″. All these quoted figures except the last need to be looked at in relation to the other factors as well; 15″ *GM* in a modern boat might be ample whereas more than 2 ft. in an older boat with low freeboard might even be critical. The problem is to combine the greatest amount of safety with qualities which will not give such stiffness that work on deck is seriously affected.

If the vessel is too tender, the crew will feel insecure and if too stiff with rolling accelerations in excess of 1 g (about 3.2 ft./sec²) they will be adversely affected in their work. A heavy deck load will influence the rolling period and decrease *GM*.

Under United Kingdom regulations [The Fishing Vessels (Safety Provisions) Rules 1975], every fishing vessel of 12 metres (39′ 4½″ approx) Registered length and over is required to satisfy a fairly specific minimum stability aggregate. This is to be applicable to all foreseeable operating conditions after due correction for the free surface effects of any liquids in tanks. The specific requirements in relation to the statical stability curve shown are given below (see Fig. 17):

A — area under curve up to 30 degrees to be not less than 0.055
    metre-radian (10.34 feet degrees).

B — area under curve up to x degrees to be not less than 0.09
    metre-radian (16.92 feet degrees).

C — area between 30 degrees and x degrees to be not less than 0.03 metre-radian (5.64 feet degrees).

x — 40 degrees or any lesser angle at which the lower edges of any openings in the hull, superstructure or deckhouses which lead below deck and cannot be closed weathertight, would be immersed.

E — maximum *GZ* to occur at angle not less than 25 degrees and to be at least 0.20 metre (0.66 foot) at an angle equal to or greater than 30 degrees.

F — initial *GM* to be not less than 0.35 metre (1.15 feet).

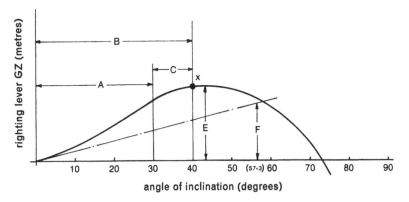

FIG. 17. Statical Stability Curve (GZ Curve).

Where vessels are engaged in single or twin boom (outrigger) fishing, virtual centres of gravity are involved (see 'Suspended Weights' pages 32 and 33) and the above values for dynamic stability, righting lever and metacentric height are required to be increased by 20%.

Some older existing vessels possess neither hydrostatic particulars nor a lines plan from which to obtain them. In such cases the owner may opt for the stability to be approved as for a new vessel (which would entail the considerable expense of drydocking or slipping to take off hull offsets for the construction of either or both a lines plan and a computer program) or by a simplified criterion of metacentric height. The formula for this criterion varies only very slightly in detail, but not in principle, as between the British Rules which cater for vessels under about 80 feet in length (or 24 metres) and the recommendations of the Maritime Safety

Committee of IMO, which apply to vessels of under 30 metres (98.5 feet) length.

The *GM* criterion in both cases is calculated from a formula which applies only to vessels having ratios of freeboard to beam and beam to depth between certain limits and of standard sheer; and it does not apply to vessels engaged in single or twin boom fishing at all. The actual *GM* is then measured by means of an inclining experiment based upon an estimated displacement or a rolling period test (q.v. Appendix 8) and its value must be at least equal to the minimum required value of the *GM* as calculated by formula. But in no case should the required *GM* be less than 0.40 metres.

The variation in metacentric height throughout a fishing trip depends upon the type of vessel. In particular, the older steam trawlers show important differences from the more modern motor ships. In steamers, the *GM* is decreasing all the time, but for motor trawlers it will start to increase again after an initial decrease soon after the first few hauls on the fishing grounds. This will continue till about half-trip condition after which it will be decreasing all the way back to the home port.

Nobody assumes that in practice all adverse heeling moments (wind, water, shift of cargo, deck loads, free surface effect, etc.) act simultaneously. It is also admittedly true that in good weather conditions a vessel will be quite safe with what might normally be considered very marginal stability. But 'getting away with it' should not bring stability instructions of a formal nature into disrepute. It is quite impractical to enter a weather factor into stability statements.

Some authorities point out that it is quite impossible to capsize a ship by resonance between the natural period of roll of the ship and regular transverse waves even if the *GM* is low. If this implies that such conditions are not dangerous then such an implication would be misleading; the shipping of green seas, breaking crests and other out-of-balance forces cannot be ignored. Heavy rolling should always be regarded as a potential hazard. For stability purposes, the effective angle of heel is the angle to the prevailing wave slope and not to the natural vertical. If, in a beam sea, a wave with a slope of 30 degrees passes under the hull when the ship's initial heel is already 30 degrees then the momentary angle of heel will be 60 degrees. Furthermore, the passage of the

TABLE I

Approximate Seagoing Stability Scale for Trawlers with Beams between 20 and 35 feet

GM ft	Classification	Period of Roll sec.	Behaviour
Under 2	Tender	6 —10	Effect of wind heeling moment most noticeable especially from abeam. Vessel tends to ship water on lee side when heeled and deckwork difficult. Crew feel apprehensive.
2—2.75	Average	$5\frac{1}{2}$—9	Easy movements, little or no seas shipped. Deckwork and trawling able to proceed normally.
3—3.75	Stiff	5 —$8\frac{1}{2}$	Rolling jerky and water being shipped on deck. Deckwork becoming difficult.
4 or over	Very stiff	4 —7	Rolling becoming violent and 'green' seas coming over windward rail. Deckwork out of the question and trawling soon suspended.

In the figures given for the rolling period the lower figures apply to the shorter beamed vessels and the upper figures to the large beams.

wave formation may produce for a brief period a reduction in the ship's displacement, and therefore in her righting moment in combined pitch and roll. Oblique seas also reduce the average stability below that calculated on statical considerations over a wide range. The most dangerous point is when the wave crest is amidships and running before a high following sea is particularly dangerous. Rolling will become most violent when the seas approach the ship from astern at an angle of about 75° (15° abaft the beam).

Trim

PREVIOUS chapters have dealt with the transverse inclinations of ships, i.e. rotation about a longitudinal axis. It will be clear, however, that inclination in other directions also takes place simultaneously, i.e. towards the bows or stern, about a transverse axis. Furthermore, these inclinations are combined in the final position of inclination. It is convenient for the purposes of analysis however, to treat these two inclinations as independent and longitudinal stability without heel as analogous to that for transverse inclinations. This is not rigorously correct, but sufficiently so for practical purposes.

Longitudinal inclination again brings in the position of the metacentre and the centres of gravity and buoyancy in the longitudinal centreline plane but the angles of inclination considered are small by comparison with those in the transverse direction. What we are more concerned with here are the practical problems arising out of changes in trim.

Trim is the difference between the draft aft and the draft forward. It is usual to express this in inches. The direction of inclination must also be given so that a trim of 30 inches by the stern means that to attain an *even keel* position the draft at the stern would have to decrease by approximately half this amount whilst the draft forward would have to increase by a similar 15 inches. This is only approximately correct because it assumes that the vessel rotates about a horizontal transverse axis amidships. In fact, this is not so because the waterplane is not symmetrical about amidships. Whilst the accuracy of apportioning the trim half forward and half aft is adequate for many rough calculations, greater precision is often required in many instances.

The point about which a ship pivots (see Fig. 18) is called the *Centre of Flotation* (or Tipping Centre) which is, in fact, the centre of area of the waterplane. Its position is located at so many feet forward or aft of the midship point. In the loaded condition, the *CF* will normally be abaft midships.

Mean draft, however, is not in practice measured at the *CF* but at midships. *Mean draft* is the sum of the forward and after drafts divided by 2.

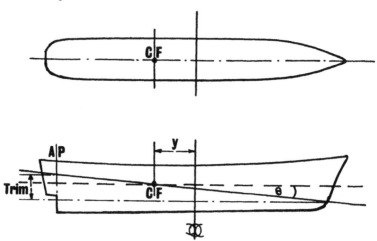

FIG. 18. Trim and Centre of Flotation.

It is unusual for a ship to be operating on even keel, i.e. zero trim. The loading of cargo and the distribution of equipment have many variations and the consumption of fuel, fresh water and stores all cause continual changes in operating draft and trim. For these reasons a knowledge of trim is essential for the ship's officer in order for him to be able to operate his ship properly. He must be familiar with trimming problems and be able accurately to estimate trim changes consequent upon the loading of cargo and the consumption or transfer of fuel or ballast.

The trimming moment which causes a change in trim of 1 inch is called the *Moment to Change Trim 1 inch* and abbreviated *MCT* 1″. Values of *MCT* 1″ (in tons ft.) are given in the builders' particulars for even keel drafts at intervals of about 1 foot. Otherwise *MCT* 1″ can be calculated from:

$$MCT\ 1'' = \frac{\Delta \times GM_L}{12L}$$

Where L is the length of the ship and GM_L is the longitudinal metacentric height (M_L is the longitudinal metacentre). A derivation of this formula can be found in Appendix 4.

Example 6

A small fishing vessel displaces 130 tons on a waterline length of 67 ft. The longitudinal metacentric height is 50 ft. The trim condition of the vessel is such that the propeller tips are 6 inches above the water surface. The after peak tank with its c.g. 26 ft. from amidships is empty. Calculate the amount of water ballast which must be transferred from the fore peak tank with its c.g. 29 ft. from amidships which is necessary to immerse the propeller tips 12 inches assuming the centre of flotation is amidships.

The change of trim required to immerse the propeller tips will cause equal changes in draft at the bow and stern if the CF is amidships, i.e. $2(6 + 12) = 36$ inches total change of trim.

Let w be the weight of water ballast to be transferred

Then, trimming moment $= w(26 + 29)$

But the trimming moment also equals $36 \times MCT\ 1''$

$$MCT\ 1'' = \frac{130 \times 50}{12 \times 67} = 8.1 \text{ tons ft.}$$

$$w(26 + 29) = 36 \times 8.1.$$

$$\therefore w = \frac{36 \times 8.1}{55} = 5.3 \text{ tons of water ballast}$$

(Note: Usually the value of $MCT\ 1''$ will be available from the builders' particulars for the draft in question. If not, BM_L being approximately equal to GM_L may be used in the formula for $MCT\ 1''$ for practical trim problems. BM_L is frequently more easily obtained than GM_L.)

If it is wished to move weights w_1, w_2, w_3 ... etc. already aboard distances of d_1, d_2, d_3, ... etc. ft. then there will be no change in mean draft. If moments are taken about the CF ($-ve$ forward; $+ve$ aft) then the algebraic sum will be the trimming moment and the $\pm$ will determine whether the ship trims by the head or by the stern. If the weights are added or deducted the procedure is the same as the above except for a change in mean draft (bodily change in draft). Then:

Bodily rise or parallel sinkage (i.e. change in mean draft)

$$= \frac{w}{TPI} \text{ (in.)}$$

$$\text{and Trim (inches)} = \pm \frac{\text{Trimming Moment}}{MCT \; 1''}$$

($+$ trim by stern; $-ve$ trim by head).

The trim is then proportioned out forward and aft according to the position of the centre of flotation, i.e. $\dfrac{CF}{L} \times \text{Trim}$

Summarising, the procedure is first to calculate the parallel (bodily) rise or sinkage of the ship as if the weight were deducted or added without trim (for all practical purposes at the *LCG*) and then to trim the ship as for a shift of weight already aboard. This does not apply to large changes in weight.

If the amounts and shifts of weights added or deducted are known including that of the ship (*LCG*) and the position of the *LCB* can also be obtained then the lever of the couple causing trim is the horizontal separation of *LCG* and *LCB*. It will be clear that as the ship trims this lever disappears so that the *LCG* and *LCB* are in the same vertical line. The ship is then in equilibrium.

$\therefore$ Trimming moment $\quad = \Delta$ (Separation of *LCG* and *LCB*)

Also Trimming moment $= MCT \; 1'' \times \text{Trim (in.)}$

$\therefore MCT \; 1'' \times \text{Trim} \quad = \Delta$ (Separation of *LCG* and *LCB*)

A refinement must be noted about reading mean drafts. It was stated earlier that mean draft was the average of the drafts at the stem and the stern, i.e. at amidships. Since a ship does not trim about amidships the mean draft of a ship with trim will not be the same as for the even keel condition at the same displacement. As displacement curves are prepared for even keel conditions, the mean draft of a trimmed ship will not correspond to the even keel drafts on the displacement scale. A layer correction must be added or deducted therefore. Fig. 19 will make this clear. A note on the "Displacement Correction for Trim" is given in Appendix 5.

A familiar practical trim problem occurs when it is desired to keep the after draft constant whilst at the same time taking weight aboard. It may be necessary to maintain propeller immersion or pass through a waterway where the depth is restricted and the

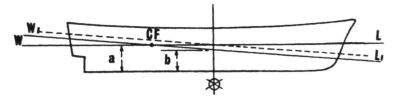

W L is the even keel waterline.
$W_1 L_1$ is the trimmed waterline.
a is mean draft before trim.
b is mean draft after trim.
Either mean draft must be corrected to even keel before reading Displacement Curve or a Displacement correction for trim must be made.

FIG. 19. Displacement Correction for Trim.

after draft must be kept constant. The limit is the moment such as will not increase the forward draft by more than the existing draft aft, i.e. even keel with no increase in maximum draft.

Consider the problem in two stages. Neglecting trim, any weight added will cause a parallel sinkage of w/TPI. If the weight is now moved a distance *d* forward of the centre of flotation the trimming moment is $w \times d$. By the requirements of the problem

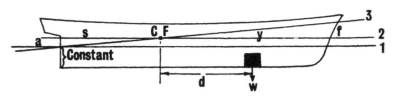

FIG. 20. Position at which weight must be added or removed to keep the After Draft constant.

this trimming moment causes an increase in draft forward and a decrease in draft aft such that the net effect of the latter and the parallel sinkage is zero change in draft aft, i.e. they cancel each other out (see Fig. 20).

L is the length of the ship.

$a + f$ is the trim (inches)

s is the distance of the CF from aft (ft.)

y is the distance of the CF from forward (ft.)

w is the added weight at distance d from the CF (tons)

$\dfrac{w}{TPI}$ is the parallel sinkage (inches)

It is a condition that $a = \dfrac{w}{TPI}$ for zero change in draft aft

By similar triangles: $\quad \dfrac{a}{a+f} = \dfrac{s}{L}$

And Trim $= a + f = \dfrac{w \times d}{MCT\,1''}$

$$\therefore a = \dfrac{w \times d}{MCT\,1''} \times \dfrac{s}{L}$$

But $a = \dfrac{w}{TPI}$

$$\therefore \dfrac{w}{TPI} = \dfrac{w \times d \times s}{MCT\,1'' \times L}$$

$$\therefore d = \dfrac{MCT\,1'' \times L}{TPI \times s}$$

If the CF is approximately amidships, i.e. $s = \dfrac{L}{2}$ then the formula becomes $d = \dfrac{2 \times MCT\,1''}{TPI}$ This, therefore, is the longitudinal position from the CF or amidships where a weight must be added to keep the after draft constant. It should be noted that this position is *independent of the amount of weight added or removed*. But w must not be too large otherwise the $MCT\,1''$ and TPI values will no longer apply.

As has been noted for transverse inclinations, the density of the water in which the ship is floating affects her trim. The change however in passing from fresh to salt water is of so little account from an operational point of view that the small trim by the stern is of little practical interest to seamen.

It might be useful to add that, at normal speeds, fishing boats tend to trim by the head and the *CG* to sink in space under smooth water conditions. This is due to negative dynamic forces on the hull between bow and stern. The effect is a little less than a foot increase in draft at the bow for medium sized trawlers in relation to the draft at rest. Therefore, a trawler on even keel in port will run with a bows-down attitude at sea.

Most craft run best with a trim by the stern and fishing vessels are normally trimmed aft in operation. The practical advice offered is that in deciding upon the amount of trim by the stern required the above effect should be taken into account and compensated by increasing the 'at rest' draft aft by an equivalent amount.

Docking and Grounding

Docking Stability and Trim

SHIPS usually enter drydock with a moderate trim by the stern. The less the trim the less the angle that the keel makes with the line of keel blocks and therefore the less water that has to be pumped out before the ship takes the blocks 'all fore-and-aft'. The ship is first centred in the dock over the line of blocks and adjusted so that there is no excessive overhang at the stem or stern.

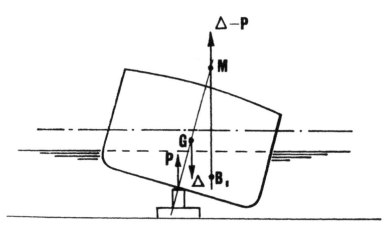

FIG. 21. Docking stability.

It is absolutely vital that the ship takes the after block properly because 'tripping' of the blocks would result in serious consequences. The after row of blocks are tied together to help counteract this. The ship should have positive GM.

As the vessel takes the after block, the upward pressure is equivalent to a negative weight of like amount at the keel. As the vessel takes the blocks all fore-and-aft, this negative weight materially reduces the effective GM. A vessel with a small initial GM could develop instability and list to an inconvenient angle. In such a case, the trim should be as little as possible so that effective GM is maintained until the ship is practically landed. Fig. 20 shows the situation.

It will be clear that as the ship is no longer fully afloat, the ship's weight Δ will be in excess of the buoyancy by an amount P which is the pressure on the blocks. This buoyancy acting through B_1 equals $\Delta - P$.

The stability condition can be investigated by considering either a virtual centre of gravity or a virtual metacentric height. As moments of statical stability have already been discussed, the latter will be taken where M_v is the virtual metacentre.

The moment of statical stability,

$$\Delta GM_v \sin \theta = \Delta GM \sin \theta - P.KM \sin \theta$$

$$= \Delta \sin \theta \left[(GM - \frac{P\ KM)}{\Delta} \right]$$

$$\therefore GM_v = GM - \frac{P\ KM}{\Delta}$$

Needless to say, GM_v must be positive for the vessel to remain upright. At the instant the vessel takes the blocks all fore-and-aft, the moment about the centre of flotation will be Px where x is the distance from the suing point to the CF. This must equal the trimming moment for even keel conditions, i.e.

$$Px = MCT\ 1'' \times \text{Trim (inches)}$$

$$\therefore P = \frac{MCT\ 1'' \times \text{Trim}}{x\ \text{(ft.)}} \quad \text{(tons)}$$

$$\text{if } CF \text{ amidships, } P = \frac{2MCT\ 1'' \times \text{Trim}}{L}$$

Also at this instant, the new displacement is $\Delta - P$ and an inspection of a displacement curve will give the draft which is the depth of water above the *top* of the blocks.

GROUNDING

Hull Pressure and Trim after Grounding

If a vessel runs aground on a rising tide she will float off of her own accord in the normal course of events provided she is not making water. If, however, the tide is falling it is expected that an alteration in trim will take place accompanied by pressure on the bottom plating. If the trim is not excessive in the first instance (when the ship runs aground), a level waterline value of *TPI* and *MCT* $1''$ can be used. Consider the unfortunate vessel in Fig. 22 having run aground forward on a rock on a falling tide. Assume a period of time in which the tide falls f ft. from W_1L_1 to W_2L_2.

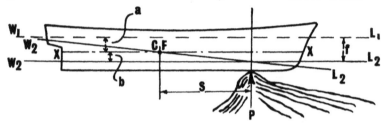

FIG. 22.   Hull Pressure and Trim due to Grounding.

W_1L_1 is the waterline at which the vessel runs aground.
W_2L_2 (dotted) is the waterline after fall in tide f ft. assuming there was no trimming.
W_2L_2 (solid) is the actual trimmed waterline.
CF is the centre of flotation.
s is distance of CF from point of grounding (ft.)
XX is a line horizontal to W_1L_1 through CF.
P is pressure on outer bottom plating (tons).

To find Pressure P

$$\text{Trim change per ft. of length } L = \frac{\text{Total Trim Change}}{L} \text{ (ft.)}$$

$$\text{Trim change in } s \text{ ft.} \quad = b$$

$$= \frac{s \times \text{Trim}}{L}$$

$$\text{but } a \quad = f - b \text{ (}a \text{ and } b \text{ in feet)}$$

$$= f - \frac{s \times \text{Trim}}{L}$$

$$P = \Delta_1 - \Delta_2 \text{ (from } W_1L_1 \text{ to } W_2L_2)$$

$$= a \times 12 \times \text{mean } TPI \text{ (between } W_1L_1 \text{ and } XX)$$

$$= 12 \times \text{(mean) } TPI \left[f - \frac{s \times \text{Trim}}{L} \right]$$

N.B. Trim in this formula is in feet. If trim is in inches then the fall in tide f must also be in consistent units.

To find Change of Trim

The longitudinal trimming (righting) moment is

$MCT\ 1'' \times \text{Trim (ft.)} \times 12$.

The longitudinal trimming (upsetting) moment balancing this is $P \times s$.

Substituting for P,

$$12 \times TPI \left[f - \frac{s \times \text{Trim}}{L} \right] \times s = MCT\ 1'' \times \text{Trim} \times 12$$

$$s \times TPI \times f - \frac{s^2 \times TPI \times \text{Trim}}{L} = MCT\ 1'' \times \text{Trim}$$

$$s \times TPI \times f = MCT\ 1'' \times \text{Trim} + \frac{s^2 \times TPI \times \text{Trim}}{L}$$

$$= \text{Trim} \left[MCT\ 1'' + \frac{s^2 \times TPI}{L} \right]$$

$$\therefore \text{Change of Trim (ft.)} = \frac{s \times TPI \times f}{MCT\ 1'' + \dfrac{s^2 \times TPI}{L}}$$

The derivations of these formulae are not particularly important, but a knowledge of the formulae themselves is of practical utility.

Example 7

A trawler grounds 51 ft. forward of midships; centre of flotation 1 ft. forward of midships. If the bottom shell plating at the point of grounding can withstand a maximum pressure of 100 tons what is the greatest fall in tide which will still leave the vessel in a safe condition? The trawler is 140 ft. in length; $TPI = 7$; $MCT\ 1'' = 55$ tons ft.

$$\text{Change of trim} = \frac{(51 - 1) \times 100}{55} = \frac{1000}{11} \text{ inches}$$

$$\text{Also, change of trim} = \frac{s \times TPI \times f}{MCT\ 1'' + \dfrac{s^2 \times TPI}{L}}$$

$$\frac{1000}{11} = \frac{50 \times 7 \times f}{55 + \dfrac{50^2 \times 7}{140}}$$

$$\therefore f = \frac{1000 \times 180}{11 \times 50 \times 7}$$

$$= 46\tfrac{3}{4} \text{ inches}$$

Dynamical Stability and Motion in a Seaway

Dynamical Stability

DYNAMICAL stability is the work done in heeling a vessel to any specified angle of inclination. In other words, it is the work done against the righting moment from upright to the angle of heel under consideration. It is a measure of a vessel's power of recovery.

Dynamical stability is of interest in investigations of transverse motion at sea, particularly the effects of sudden squalls and also rolling. Dynamical stability is measured by the area under the statical stability curve. Integration of a curve of statical stability from upright to any angle of heel θ is equal to $\Delta GZd\theta$ which is the work done or dynamical stability. If the ordinates of the statical curve are in GZ ft., then it is of course necessary to multiply the the result by the displacement Δ for work done in ft. tons. Integration of a statical curve for successive angles of heel will provide ordinates for a new curve of dynamical stability, but this is not often required in practice. For the usual integration rules the common interval between angles of heel must be converted to radian measure.

Power to Carry Sail

The effect of wind-heeling moments on ships is amply illustrated by considering a sailing fishing vessel under normal seagoing conditions. The following assumptions are made:

1. At any angle of heel the sail area upon which the wind acts is the vertical projection or elevation of the actual area. The sails are assumed to be trimmed flat.

2. The Centre of Effort (*CE*) is at the *CG* of the sail area.

3. The Centre of Lateral Resistance (*CLR*) of the boat is the centroid of the immersed lateral plane (the underwater hull shape as seen from abeam).

4. The projected arm of the heeling moment is the vertical distance between the vertical *CE* and the upright position of the *CLR* × Cosine of angle of heel.

The following notation applies:

A = actual sail area (sq. ft.)
h = vertical distance *CLR* to *CE* (ft)
p = wind pressure (lb./sq. ft.)
θ = angle of heel (degrees)

(The *CLR* may be assumed to be at mid-draft).

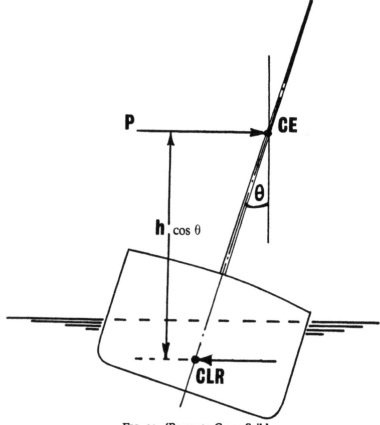

Fig. 23. 'Power to Carry Sail.'

Then,
Horizontal projection of (or effective) sail area is $A \cos \theta$
Horizontal wind force $= pA \cos \theta$
The lever arm (Vertical projection of h) $= h \cos \theta$
Righting Moment $\Delta GZ = \Delta GM \sin \theta$ (ft. tons)
Heeling (upsetting) Moment $= pA \cos \theta \, h \cos \theta$
$= pAh \cos^2 \theta$ (lb.)

Therefore at steady heel (i.e. equilibrium)

$$\Delta GM \sin \theta = \frac{pAh \cos^2 \theta}{2240}$$

$$\therefore A = \frac{\Delta GM \sin \theta \times 2240}{ph \cos^2 \theta}$$

This equation is known as the 'power to carry sail' under any assumed wind pressure. If A and p are constant and for small angles $\sin \theta$ and $\cos \theta$ are approximately equal to one, then $\Delta GM \times 2240 \div h$ is also a constant factor at a steady angle of heel. The assumptions given are not perfectly true but the above theory serves as a first approximation applying to all surface displacement vessels whether carrying sail or not and is of use in the section on 'Wind and Water Moments'. In the case of sailing ships themselves, however, the theory is of comparative use only (which can still be very useful) because the assumptions are seriously in error when applied to cloth sails and sailing ship hull forms. In practice, the wind force is not proportional to the projected area, does not act exactly at the centre of area nor in the direction of the wind.

The conventional assumption that the Centre of Effort is at the geometric centre of the working sail plan is far from true in practice. Sails cannot be trimmed flat even when beating to windward and for other points of sailing the sheets are 'started' (slackened off) and the *CE* then swings off to leeward. Also, the sails have an influence upon each other which affects the *CE*. In the close-hauled condition, the actual *CE* of a sail is approximately 20 per cent of the sail's mean width abaft the sail's leading edge (the luff) which is much further forward than the geometric centre.

Similarly, the Centre of Lateral Resistance is not at the geometric centre of the ship's underwater longitudinal plane. This is not true even when the ship is upright and its position varies with angle of heel, leeway, speed and fullness of hull form either forward or aft of the initial position.

The foregoing remarks do not entirely rule out the simple basic theory as a practical tool. Rather are they intended to point out its lack of validity without considerable refinement. One can be assured that the conventionally derived relationships between *CE* and *CLR* are still of considerable practical utility, especially in experienced hands, when comparisons between one sailing ship design and another of known performance and hull-sail balance are required rather than actual parameters.

The refinement of the theory of the mutual relationship of *CE* and *CLR* both vertically and horizontally is one of the most crucial and fundamental aspects of the design of all sailing ships because it affects the whole balance (controllability) and performance of the vessel. Further treatment of the criteria, including stability, is extremely complicated and of interest mainly to designers and hydrodynamicists.

SAILING SHIPS IN PRACTICE AND POWER-SAILING

(or the Combination of Sail with an Auxiliary Source of Power)

One way of deciding what working sail area should be carried is to take the optimum sailing angle of heel for the type of vessel in question at an acceptable wind force (which might vary perhaps from 15 to 20 knots) and then to calculate the permissible sail area to fulfil the conditions. Optimum sailing angles are, of course, more concerned with least resistance than with safe stability but the stability must be suitable for the sail area carried, neither too stiff nor too tender. Whilst the former is not usually a problem in this connection, a tender vessel on the other hand would have to reef too soon and, therefore, lack the necessary drive in the very conditions which had forced her to reef. The amount of sail which can be carried in any particular wind and sea conditions varies according to how the vessel is loaded because this affects the detailed stability characteristics and therefore the degree of stiffness applying at the time.

However, the traditional way of determining the amount of sail area to be spread on sailing ships is mainly by comparison with other vessels whose performance is known and comparing ratios of sail area to wetted surface and the $\frac{2}{3}$ power of the displacement in conjunction with various hull form proportions.

At one time, there was a very large number of different types of commercial sailing fishing vessel throughout the world. Now, with a few isolated exceptions, they seem to have been relegated to the maritime museum and the pages of the history books. However, one cannot but wonder in an era of scarce and costly fossil fuels and efforts to conserve them, whether this was not a little premature. Resurgence of interest in sail is an established fact with experimental commercial sailing vessels at sea now. It can by no means be discounted that such a reversion (or economic development) will not embrace some small modern fishing vessels some of whose hull forms are quite suitable for modest sailing rigs as an auxiliary source of power. Indeed, power-sailing can be not only most economical but also the most comfortable way known of working a small vessel up to windward. As an example, and according to Capt. S. Remøy of the Government School of Fisheries in Norway the Arctic seal-hunters still use sails to this day even though they have powerful engines. The ketch rig consists of fore staysail, 'three-hooked' mainsail on the foremast and a mizzen. All are sheeted close-hauled to keep the ship up to wind and sea and the engine is used only sporadically to bring the bow up again should it fall off. The system works well going against heavy weather and fuel consumption is minimised; moreover, when drifting freely, rolling is significantly reduced by the damping effect of the inexpensive 'three-hooked' mainsail kept tight by tacks amidships. In long and heavy rolls it is claimed that it is possible to increase work on deck sometimes by as much as 50% using this steadying canvas.

Whilst it is not the function of this book to delve into the design aspects of putting sailing rigs back onto fishing hulls, some understanding of the effects of sail on handling and stability, including directional stability, might not be out of place in anticipation of possible future circumstances.

Most middle-distance British fishing smacks were ketches with a few sloops. The trawl was always towed with the warp sheeted over the *windward* quarter. This was most important in case the trawl snagged on the bottom as will now be explained. Basically, there were two working hatches in the deck, a fish hatch and a warp hatch, the latter always being the forward one. The end of the warp was made fast in the hold to the bottom of the mainmast and payed out of the hatch with turns taken round an adjacent capstan or winch on deck. The warp then usually passed over a roller on the weather rail opposite the capstan and then back aboard to the pumphead amidships where it was secured by a stopper (sometimes referred to by the fishermen as 'legs') consisting of about 2½ fathoms of plaited rope. The warp then led directly over the weather bulwark against a thole pin into the sea and the bridle of the trawl. A guy consisting of wire and then rope led from a winch forward of the mainmast to a stemhead roller over which it passed back along the weather bulwark, all outside of the shrouds, to the bight of trawl warp between the capstan and the pumphead.

This somewhat lengthy description of what is quite a simple arrangement is necessary in order to explain how the old sailing trawlers avoided gybeing or being brought up 'all standing' if the trawl caught on the bottom. It also avoided the potentially awkward stability problem so familiar to all tugmasters when they get a towline leading abeam (*ie* the danger of being over-rolled). With such an arrangement of trawl warp as described above, the smack would just luff head to wind thus giving the crew ample time to sort matters out. The rope stopper on the pumphead was the self-rendering overload protection of those days—cheap and effective. Similar conditions would apply to any present day fishing vessel relying for part of its propulsion on any fore-and-aft sailing rig.

As every seaman knows, sailing like navigation, is an art only acquired by practice and experience; but some understanding of the theory, some idea of what one is doing and why, is essential in improving personal performance notwithstanding any natural aptitude. Any opposite view is contradicted by the large number of volumes on both subjects which have been making their way onto and off the library shelves since long before the printing press was invented.

Wind and Water Moments

The angle of steady heel can be found by drawing curves of wind heeling moment and statical stability on the same base (i.e. angles of heel). The curves will intersect at the angle of steady heel.

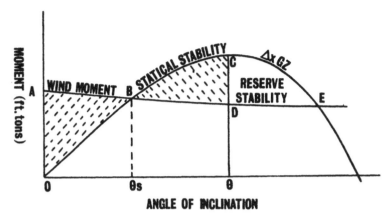

FIG. 24. Wind Heeling Moment.

This can, of course, be applied to any vessel apart from sailing ships. Suppose now a motor trawler in the upright position were exposed to a sudden squall directly abeam. The work done by the wind will be the area under the heeling (upsetting) moment curve. The work done by the dynamical stability of the ship is the area under the stability curve. In Fig. 24 the ship will heel to some angle θ such that area *AOB* equals *BCD*. At this angle the work done by the wind equals the work done in overcoming the ship's resistance to heeling. If the wind remains steady the ship will settle back to a steady angle θs where the upsetting and righting moments are equal. The area *CDE* is the reserve of dynamical stability. At any point up to θs the ship must be considered safe. At any point beyond *E* the ship will capsize. If any upsetting moment in excess of Δ*GZ* max is applied steadily the ship is in danger and may also capsize.

Let us now take this case a stage further. The trawler is again exposed to a sudden squall but this time she is rolling freely. Just as she has completed a roll to windward the squall strikes on the vessel's windward side. Referring to Fig. 25 the statical curve is *OY*. The righting moment *XY*, returns the vessel to the upright at the same time as the wind moment *XA* acts. The total moment now becomes *AY* (i.e. considerably increased). This is sufficient to carry the vessel over to an inclination θ and graphically illustrates

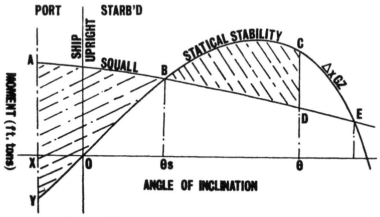

FIG. 25. Effect of sudden squall on ship when Rolling.

the danger to which a small vessel with small maximum *GZ* and short range stability (i.e. inadequate reserve dynamical stability) may be exposed.

In any realistic consideration of the stability of ships in a dynamic environment one must also take account of shipping water. Most fishing vessels are fitted with bulwarks for practical reasons and after any roll towards a wave this arrangement is capable of holding water. It is important to understand that this water has an influence on the effective *GM* due to its weight and height of centre of gravity and free surface. This will apply if the side scuppers and freeing ports are inadequate to clear the deck before the vessel again returns to ship the next wave. Normally the freeing arrangements could not be expected to cope and, indeed side rails would not completely eliminate this effect. Fig. 25 shows the contained water moment acting against the stability.

Normally, shipping water by itself is not a serious matter provided quantities are not allowed to flood through deck openings and freeing ports are adequate. During fishing operations, progressive flooding of fishrooms could occur through hatches being open and not being capable of rapid closing. The Torremolinos International Convention for the Safety of Fishing Vessels, 1977 recommended that the angle of heel at which this could happen should be at least 20 degrees unless the values of what now amounts in most cases to statutory stability criteria can be met with such holds partially or completely flooded. It is when a combination of broadside-on shipment of heavy 'green' seas, large angle rolling and wind act in unison that critical conditions can

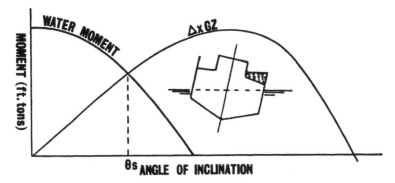

FIG. 26. Effect of shipping 'Green' sea.

occur. This would most particularly be so if the ship were already broached-to and held steady at some angle of heel on the crest of a wave when stability is effectively reduced. This is highly complex

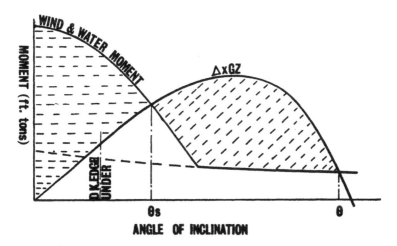

FIG. 27. Critical Stability Situation (Ship would probably capsize).

and beyond the scope of this book, but suffice it to say that a cycle of such conditions would probably prove fatal to a small vessel. On the other hand, it would probably be a case of bad seamanship and mishandling to get into such a situation in the first place. It is nevertheless important to know what might happen.

Fig. 27 shows the combined wind and water effects. Such conditions may obtain in northern waters in winter. A baulked roll would probably occur at θs.

Figures 24, 25, 26, and 27 are attempts to show, graphically, a simplified picture of the credit forces versus the debit forces in the stability balance, but because the subject *is* complicated they are to some extent theoretical. However, it is most important not to dismiss them as a mere theoretical exercise; to do so would be like throwing the baby out with the bath water.

For those who wish to go more deeply into the problems, the simplifications involve treating the dynamic scenario as if it were a static or quasi-static one. As discussed in the next section on motion, rolling and wave action are dynamic, time-based quantities and the graphs omit any time scale. Nor is it rigorously correct in the mathematical sense to relate the rolling motion of a ship to wind and wave action separately and then to sum their effects algebraically in order to arrive at their simultaneous effect. In bad weather, winds do not blow steadily; they gust. By 'gustiness' we mean that the wind blows erratically with increased force for short periods of time and that cannot properly be related to a vessel's rolling angle so it cannot be related to a *GZ* curve either. The vessel in Fig. 23 is not, moreover, in static equilibrium under a steady wind force balanced by an equal and opposite water resistance force because, in fact, practical experience shows that vessels suffer from drift and leeway. There still seems to be uncertainty about which point one should take moments of wind heeling force so the length of the lever arm is in some doubt; apart, that is, from determining the centre of pressure.

In Fig. 26 certain effects of water on deck are ignored. Additional time-related effects are the kinetic energy of the water (surge) and its transverse sliding (of course the ship is pitching too). When the rolling of the ship and the water motion are in-phase this is like the effect of a water-chamber stabiliser (anti-rolling) tank acting the wrong way—unstabilising the ship! But in spite of all this, it is repeated with emphasis that the graphs are still worthwhile.

To withstand bad weather, fishing vessels should pay particular attention to the following points:

1. The ship shall be initially upright and swing about this position.
2. That the *range* of stability should be large. (This is quite a useful criterion of stability.)
3. No sizeable opening can be submerged by a large angle of heel. (This involves closing arrangements being proof against heavy weather damage.)

MOTION AND SEAKINDLINESS

A vessel may be absolutely *seaworthy* and yet be extremely uncomfortable. Seaworthiness is a quality difficult to define in precise terms but naval architects and practical seamen can usually be found to possess strong views about unseaworthiness. Controversy usually centres round ship motions and behaviour in a seaway. The only point at issue is that of safety.

Seakindliness on the other hand is a different matter. A seakindly vessel behaves in a manner which makes the exercise of expert shiphandling minimal even though such expertise is expected to be instantly and always available. The point at issue here is one of comfort. No heavy rolling, no excessive accelerations, oscillations of small amplitude, good steering response, freedom from spray and green seas and so on. These various motions in a seaway will now be examined. There are six of them: 1) rolling; 2) pitching; 3) yawing; 4) heaving; 5) surging and 6) swaying (drifting or leeway). All six of these motions are probably taking place at the same time although rolling or pitching will most likely predominate. However, it is usual to analyse only one of these motions at a time.* Indeed, it would be an almost impossible task to attempt to do otherwise. The first three motions are oscillations about some axis; the latter three are translational. Some appreciation of ship motions is important to the practical seaman if only to impress

---

*The mathematical treatment of these in any detail is extremely complicated and beyond the scope of this book.

upon him that the statics of stability and trim are comparative assessments and very useful tools, but not necessarily statements of what actually happens in practice—in a dynamic environment.

Rolling

Rolling about a longitudinal axis is usually considered to be the most important periodic motion of a ship at sea. It affects the personal comfort of everyone aboard and to a large extent conditions any work on deck. It is thus of great importance in fishing vessels of all types. Rolling is related to the stability of the ship and therefore her safety at sea. Resistance to rolling and the vessel's rolling period are aspects of primary importance.

A ship does not behave like a pendulum with its point of suspension at M and the CG acting as the bob. A ship has no fixed axis of oscillation but what is called the *instantaneous axis* is located somewhere near the CG. Of course, the CG itself describes a path in space as the vessel rolls even though, relative to the ship, it remains fixed. The period of roll given below differs from that of an 'equivalent' pendulum because the ship has a varying moment with angle of heel, in addition to that due to gravity, which a free pendulum lacks.

A period of single roll is the time taken to roll from port to starboard, i.e. from 'out-to-out'. If a ship were made to roll in still water, the rolling periods would all be about the same (isochronous), but the angles of roll would become less and less until they were finally damped out. In other words, the amplitudes would diminish.

A complete *period of roll* is from port to port or starboard to starboard, i.e. from 'out-to-out' and back again. For unresisted rolling in still water, this is known as the ship's *natural period of roll*. The formula for this is:

$$2T = 2\pi \frac{k}{\sqrt{g \times GM}} = \frac{1.108\,k}{\sqrt{GM}} \text{ seconds}$$

T is in seconds; $\pi = 3.142$ and g, the acceleration due to gravity is 32.2 ft./sec./sec. The symbol k is called the transverse radius of gyration, which depends on the distribution of all weight in the ship.

From an inspection of the formula it will be seen that the bigger the GM the shorter will be the rolling period. The shorter the period the more the tendency to roll and the quicker it will be and therefore the greater the discomfort. The only way to get over this is to reduce *GM* and increase *k* by 'winging' any hull weights that can conveniently be moved. Vertical movements of weight will be found to have much more influence than transverse movements on the rolling period. Rolling is not isochronous for very small metacentric heights or large angles of heel (beyond about 15°).

A ship's still water natural period of roll, although of no practical utility in itself, is of great comparative value in predicting the probable behaviour of the vessel at sea (see "Determination of *GM* from Period of Roll", later in this Chapter and Table I).

Resistance to rolling is offered by:

1. Fluid friction against the hull.

2. Wavemaking of the ship.

3. The ship's speed (if the ship's hull form is not too full-bodied. Most fishing vessels have a medium to fine hull form).

4. Hull shape, especially at the turn of bilge, and appendages like bilge keels.

5. Anti-rolling devices (stabilisers).

Up to a point, a vessel in waves will behave as part of the wave. In other words, the ship will try to behave as the particles of the wave behave. This behaviour will depend upon the part of the wave in which the particles happen to be situated both as regards crest or trough and as regards depth. Of course, a ship must be regarded as rather a large particle, but certain generalities prevail, the most important of which is the ship's beam/draft ratio. Fig. 28 shows the two situations for shallow and deep drafts respectively.

The natural rolling period of the ship affects this situation. A short period will tend to make the vessel behave as in Fig. 28 (top).

If the wave period is also short the resulting motion will be most uncomfortable. The angular acceleration on deck will be rapid and the check to the roll even violent so that it may be impossible to work on deck and, at worst, damage to fittings and even local structure may result.

With a ship's natural period in excess of the wave period the motion will be easier. There is a time lag so that wave action tends to resist the ship's inclination. This is important in vessels where the crew has to work on deck as in trawlers and fish factory ships.

It was observed previously that if a ship is caused to roll in still water the rolling would eventually die out. This fact is easily demonstrated by practical experiment. Ocean waves cause the ship to roll and if these waves are regular they will, in due course,

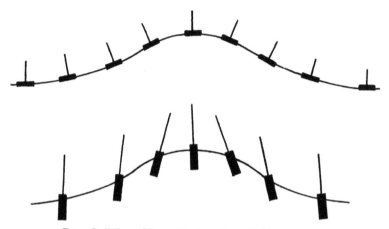

FIG. 28. Effect of Beam/Draft ratio on Rolling in waves.

impose their own periodicity on the ship instead of the ship's own natural period. This is called *forced rolling*. If the period of the waves is not constant, the rolling period of the ship will not be constant as the ship is always tending to revert to its own natural period. This is, of course, the usual state of affairs.

Trouble arises when the natural rolling period of the ship (2T) equals that of the wave. The passage of each wave adds to the inclination of the ship which builds up in magnitude until, in theory, she eventually capsizes. This is known as *synchronism*, but it should be borne in mind that a uniform increase in ship inclination can occur only in an unresisting fluid. This is, therefore, something which cannot exactly happen in practice, but a close approach thereto is perfectly possible and could be dangerous. This can be demonstrated by suitable models in a tank which can be upset by waves having periods equal to their own. The remedy for synchronism is to change course (which is much more effective than a change in speed).

If the ratio of ship period to wave period is less than 1 (which is synchronism) the ship will heel *away* from the wave crest (Fig. 28 top)) and if the ratio is greater than 1 will heel *towards* the crest. This is a point worth bearing in mind. Trawlers usually have small natural periods of roll and they will normally heel away from the

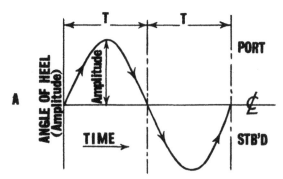

One Complete Unresisted Roll in Still Water.

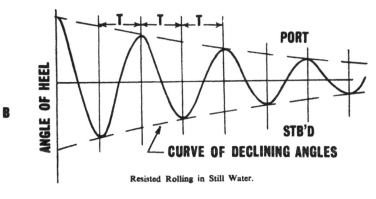

Resisted Rolling in Still Water.

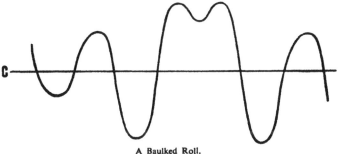

A Baulked Roll.

FIG. 29. Rolling Motions.

wave crests. However, in the very short seas not infrequently found in the North Sea, they can sometimes heel towards the crests. This causes unpleasant rolling motions.

A baulked roll is one in which the ship starts a roll from her maximum heeled angle on one side but does not complete the roll to her maximum heeled angle on the other side. Instead, the ship stops short in her swing and returns to the maximum angle on the side from which the baulked roll started. She will then go on to several forced rolls and repeat the process. This is a common occurrence at sea, but baulked rolling can be unpleasant.

If the ship has negative metacentric height or the centre of gravity is unusually high, a roll can develop into a *lurch*. This is both unpleasant and dangerous in any sort of rough sea condition. Lurches can be violent and may cause the vessel to roll irregularly into the waves with any of the expected consequences. Slack tanks can also contribute to a lurch through surge of their contents, but for surging to be of consequence the amount of liquid must be relatively large in relation to the vessel (e.g. a tanker). So surging is not usually of importance in fishing vessels unless of course a large amount of free water has been shipped.

It is helpful to be able to picture rolling motion graphically. Fig. 29a, b, c, has been produced to illustrate simple rolling diagrammatically.

Determination of GM from Period of Roll

The necessary time for an inclining test (Chapter 3) on a trawler in port cannot often be spared. An inclining test requires all other activities to cease whilst it is in progress and this is usually in conflict with important commercial considerations.

The metacentric height can, however, be found from the natural period of roll if this can be timed accurately. A fairly large number of timings must be taken and averaged if the period is to be reliably estimated.

A ship's natural period of roll is given as:

$$2T = \frac{1.108\,k}{\sqrt{GM}} \text{ or } GM = \left[\frac{1.108\,k}{2T}\right]^2$$

where k is the radius of gyration. If mB is substituted for k in this formula where B is the ship's breadth in ft. and m is a constant, it has been found that the *m*-values bear a constant average relationship to loading conditions in fishing vessels, i.e. from about 0.32 to 0.45 for steel ships and from about 0.44 to 0.51 for wooden vessels.

For steel trawlers on a normal fishing trip of about three weeks, the m-values can be taken as follows (the table is due to W. Mockel)

Departure from port	m = 0.400
Outward bound	0.395
On the fishing grounds	0.390
Homeward bound	0.387
Arrival at port	0.385

$$\text{Then } GM = \left[\frac{1.108\ mB}{2T}\right]^2$$

where $2T$ is the full period of natural roll from out-to-out and *back again* (sometimes this is just designated T which can cause misunderstanding.) If the rolling periods are taken at sea, it is important to be able to recognise forced oscillations. These must be disregarded. IMO recommendations on the Rolling Period Test are given in Appendix 8.

Pitching

Pitching is the periodic rotating motion of a ship about a horizontal transverse axis. Strictly speaking, pitching is the bows-down inclination whilst the reverse (bows-up) is known as 'scending. Next to rolling, pitching is the second most important motion of a ship at sea.

Pitching is heaviest when the ship's centreline is perpendicular to the direction of propagation of the waves and conditions of synchronism occur. This is not usually a problem because damping effects are large and any alteration of course or speed affects the motion quickly.

If the wave lengths are long and the ship speed slow (or in a following sea which is somewhat less than the ship's speed) the ship will set herself along the effective wave slope and forced pitching will take place. This forced pitching will have a period equal to the apparent period of encounter of the waves. The maximum pitching angle will not be greater than the wave slope. If ship's speed is high and the wave lengths short (a short sea) pitching amplitudes will be small. The matter which determines the wetness forward is the angle of the ship relative to the wave slope at encounter rather than the maximum pitch angle (which may occur at some less important part of the motion). A ship with a small pitching motion may therefore be wetter than one in which the angular motion is greater.

Natural pitching periods have been found to vary between $\frac{1}{3}$ and $\frac{2}{3}$ of the natural rolling periods. If ocean waves were regular, ships would generally find themselves in a forced pitching situation. Unlike rolling, this is much less so and ships mostly pitch in their own natural period.

Summing up, pitching will be affected by:

1. Ship's natural period.
2. Wave period of encounter (wave length).
3. Wave height.
4. Ship's speed (related to 2).
5. Longitudinal distribution of weights (stores, fuel, cargo).
6. Fullness of hull form.

Yawing

Yaw is the tendency for a ship to veer off course. Most ships are directionally unstable requiring the constant use of helm. Unlike the motions previously discussed, there is no restoring moment. This must be applied externally by means of a rudder (which initially causes more yaw, known as a drift angle) or some such device as a lateral thrust unit. Yawing is concerned with a vessel's steering characteristics which latter are beyond the scope of this book. Forced yawing is associated with rolling and is a factor in vessels with too little deadwood (fin area aft) and full bodied stern lines. Directional stability is also affected by windage forward.

The only control the seaman has over directional instability (a tendency to yaw) is the judicious use of helm and (in twin screw ships) adjustment of engine revolutions. An increase of trim by the stern would also help.

Heaving

Heaving (and its reverse movement *dipping*) is the vertical bodily movement of the whole ship upwards and downwards. It is caused by changes in buoyancy as each wave passes the ship. Heaving is associated with rolling and pitching and is periodic. The more a ship rolls and pitches the greater the amplitude of heave to be expected.

Surging or Surfing

Surging is the tendency of a vessel to move forward on the crest of a wave and backwards in the trough or vice-versa. This is due to the motions of the water in different parts of the wave. Naturally, a ship steaming through a swell will not actually move backwards but her motion will be alternately accelerated and retarded according to which direction she is proceeding in relation to the swell.

Surging is really only noticeable in very small slow-speed ships or in synchronous conditions. It can, however, be assisted if the screw pitches out of the water causing periodic unevenness of thrust.

Sway and Drift

When a vessel travels parallel to the waves surging action is, of course, still present. The vessel will be alternately on the crest and in the trough of the wave. A transverse movement or surge of the ship therefore takes place. Transverse surge is called sway.

As a ship rolls broadside on to waves some portion of the wave energy will be expended against the ship's side. In other words, the ship is subjected to a series of pushes. This causes drift. Usually a ship with no way on her will set herself broadside on to a swell. Of course, the wind also affects drift and the arrangement of the vessel's superstructures will affect the way in which the vessel lies to the weather. The familiar mizzen or jigger sail on drifters is an example of this. Such a device is employed to bring the ship's head up into the wind. The vessel will still drift, but rather more slowly, to leeward stern first.

Panting, pounding and slamming are impulses or rapid vibrations at the fore end of the ship due to large hydrodynamic pressure changes caused by the motion of the vessel through waves. Their technical explanation is of more interest to the naval architect than the seaman. However, unless they are also to become of interest to the ship repairer through structural fatigue their occurrence should be avoided, mainly by a reduction of speed. It is erroneous to assume that these effects are only associated with the vessel's forefoot emerging from the water; but pitching has much to do with these phenomena. Most fishing vessels have forms and scantlings which are not very susceptible to these effects in any case.

The sort of pitching situation to avoid if possible!

The illustration above shows a trawler model with a typical ice formation when head to wind. The estimated weight of ice for this particular trawler at this stage is equivalent to 140 tons.

Photo: *Fishing Boats of the World 2*   Fishing News Books Ltd.

Typical ice formation with the model stern to wind, the estimated ice weight being equivalent to 90 tons. The foremast and its associated rigging were largely blanketed by the superstructure. The loss of stability in terms of GM in this position is only about half of that in the ahead due to the lower centre of gravity of the ice.

Photo: *Fishing Boats of the World 2*   Fishing News Books Ltd.

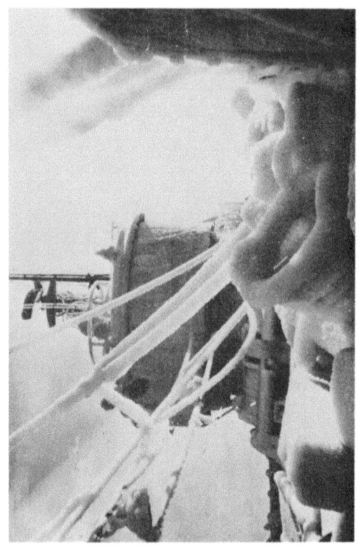

Heavy ice accumulation on fittings and superstructure of a trawler.

Stability after Damage

D AMAGE below the waterline whereby the sea floods into one or more compartments, i.e. bilging, can often result in extremely complex and serious stability conditions.

Foundering as a result of continuous or successive flooding of compartments sometimes takes a long time. If the ship sinks merely through 'letting the water in' then immediate safety of life is not usually in jeopardy. However, if an unstable situation develops it will usually do so very rapidly. If a ship sinks through instability then, instead of the ship getting lower and lower in the water until she finally disappears, the action is a capsize. The whole process may take *less than five minutes*.

Damage will result in an increase in displacement, draft and a reduction of freeboard. There will usually be list and trim and a free surface effect. All of these affect the metacentric height.

The amount of water which can enter a compartment depends upon whether it is empty or contains cargo or machinery. The percentage volume by which a compartment has been filled is known as its *permeability*. The permeability of a hold which is half-full is therefore 50%. The permeability of an empty ballast tank is 100% and when full is zero because, of course, no sea-water can enter a full ballast tank if it is torn open. Enginerooms may have a permeability of about 85%. Permeability is designated by the Greek letter μ.

The simplest case of underwater damage is the flooding of a central compartment. If the flooding is complete there is no free surface effect and no trim or list. The weight added, therefore, is the volume of the compartment multiplied by its permeability and divided by 35 (35 cu. ft. of seawater = 1 ton). The probable result is that the ship's centre of gravity will be lowered and the increased draft may raise or lower the metacentre. The net effect is nearly always an *increase* in *GM* and, in principle, is not dissimilar from the filling of a deep water ballast tank. The reduced freeboard will shorten the stability range, but this will only be serious if the remaining freeboard is inadequate to cope with prevailing weather conditions.

There are alternative ways of considering damage which results in flooding. One is to treat the flood water as *added weight*; the other is to base the investigation on *lost buoyancy*.

In the 'added weight' method the procedure is similar to that described in Chapter 2 and new centres of gravity and buoyancy determined together with trim and/or list. The new draft will depend upon the interior level of flooding assumed. Flooding will, in fact, continue until a state of equilibrium has been reached and the outside draft and interior flood level are equal. This is successive approximation as added weight increases the draft which in turn increases the amount of flood water which can be admitted, which again increases the draft and so on until equilibrium. Stability calculations assume an intact hull on the basis of added weight of water involving, of course, its appropriate free surface correction. The inertia of the free surface should be modified by its permeability (surface permeability), i.e. $I \times \mu$.

The 'lost buoyancy' method assumes, in effect, that the damaged compartment is no longer a part of the ship—it has been 'lost' to the sea surrounding the ship by virtue of the degree of free communication. The method takes the final condition when flooding has stopped and the ship is in equilibrium. The buoyancy lost must be compensated by parallel sinkage of the ship to some new waterline where the lost buoyancy is regained. The displacement remains the same (because the part of the ship 'lost' has been compensated by increased draft of the remainder). The *CG* and *LCB* are in the same position but the *VCB* rises. The inertia of the waterplane alters because only the intact portion may be considered. The value *BM* will therefore change. But there is no free surface correction to consider. Trimming will take place according to the redistribution of buoyancy (or displacement) but the amount is unaltered.

An important point in comparing the two methods from a stability point of view is the markedly different values for metacentric height. The statical stability moment ΔGZ must be the same in both cases and equal to $GM \sin \theta$. But Δ increases in the

added weight method by $\dfrac{V \times \mu,}{35}$ whereas for lost buoyancy Δ is

constant. Therefore, in order to keep the product $\Delta GM \sin \theta$ constant, the metacentric height by the lost buoyancy method always comes out greater than that by the added weight method in the inverse ratio of the two displacements.

GM is always changing (as well as Δ) if one considers flooding as added weight, but as Δ remains constant for lost buoyancy, the GM for any flooded condition is a correct and direct measure of the initial stability in the final condition. The added weight method is useful for the investigation of intermediate stability conditions. But in all flooding cases the method of determination requires to be stated in order that the appropriate displacement can be recognised.

Cause of List

It is important to determine the cause of list. This may be difficult in a damaged ship. The list may be due to instability or it may be due to unsymmetrical intact buoyancy which has the same effect as an off-centre weight (see Chapter 2).

The latter causes 'ordinary' list which may be corrected by moving weight over to the high side. If, however, the cause of list is due to instability and the ship has a loll either side of the upright (i.e. negative GM in the upright position) the shift of any weight transversely is obviously a liability to the ship. It will only *increase the list* on the other side. The remedy is to proceed cautiously and reduce top-hamper (or increase 'bottom' weight) following the listed recommendations in Chapter 2.

Stability at List

An approximate appreciation of the stability situation of a damaged ship can be obtained in certain circumstances.

If a statical stability curve is available for some standard condition of loading which is not appreciably different from the ship's condition when damage occurs, then this curve can be quickly and approximately corrected in a very simple manner.

It is assumed that the ship is damaged below the waterline and the flooding is unsymmetrical causing a list to one side. The heeling (upsetting) moment will be the weight of flood water multiplied by the horizontal lever arm d from the CG, i.e. the moment of the weight about G which is $w \times d$. (Assume G as for the intact ship.)

This upsetting moment must be balanced by the ship's righting moment ΔGZ. If necessary, alter the vertical scale GZ to read ΔGZ and then draw a horizontal line (parallel to the angles of inclination at the base) to represent the moment $w \times d$. This will cut the statical stability curve at the angle of list where the heeling and righting moments are equal. The part of the stability curve above the horizontal line drawn for $w \times d$ is the new statical stability curve.

This new curve is only approximate because, of course, no account is taken of increased Δ, the change in VCG, free surface loss or alteration in lever arm d with inclination; but these effects tend to cancel at moderate angles.

It may be stated therefore, that if the angle of list can easily be ascertained, an approximate appreciation of statical stability may be obtained within a few minutes. There is no need to evaluate the moment $w \times d$.

It will be noticed further that, as the horizontal line representing the upsetting moment $w \times d$ is raised, the reserve of dynamical stability gets less and less until $w \times d$ becomes tangential to the apex of the stability curve at GZ_{max} or ΔGZ_{max}. At this point the angle of list is critical.

Therefore, *a steady list at the angle corresponding to maximum GZ means that the vessel is unsafe and in danger of capsizing.*

Stability at Damaged Trim

Excessive trim in a damaged condition can also be dangerous especially if the flood water extends over the freeboard deck (the deck to which watertight bulkheads are carried, usually the weather deck in fishing vessels). In practice, free surface effects would certainly come into play or the equivalent loss of waterplane moment of inertia. This may extend for some distance abaft the damage due to wave action. Therefore, the effect of trim on stability is particularly onerous if water is allowed to flood extensively over an intact deck or weather deck. When damage occurs at one end of the ship it is important to confine the flood water and *prevent its spreading amidships both above and below decks.*

ICE FORMATION

Ice formation on trawlers in Arctic waters is a serious problem and the loss of vessels has been attributed to this as a primary cause.

As a result, investigations were carried out to determine the weight and distribution of ice and its effect on stability for different rigging arrangements and attitudes of ship to the weather.

A not unexpected conclusion was that the quantity of standing rigging had a major influence on both the amount of ice accretion and the height of its centre of gravity. Self-staying or tripod masts are therefore much to be preferred from this point of view, especially for head into wind attitudes. The loss of *GM* may be expected to be only about two-thirds that for normal rigging.

Other important conclusions of this investigation, which was carried out by Messrs. Vickers-Armstrongs (Aircraft) Ltd. on behalf of the BSRA, were:

The head to wind attitude results in the greatest ice formation.
The stern to wind attitude results in the least ice formation.
The loss of *GM* in the head to wind condition is about 50% greater than in the stern to wind condition.

In a head-up attitude it is likely that the ice formation will be very considerable in extent and that about $\frac{1}{3}$ of its weight will be concentrated on masts, rigging and upperworks generally. This can have a serious effect on stability lever, freeboard and therefore range of stability. With the wind fairly fine on the bow the situation is little different and the ship may list into the wind. This only makes matters worse as it appears to assist the formation of icing high up and a critical condition is reached at which the wind heeling moment is sufficient to capsize the vessel.

Currently, there is no satisfactory solution to this problem other than to put stern to wind and withdraw from the area. Even with a stern-on attitude it is only a matter of time before a critical situation develops but the amount of freeboard will have a most important influence on this.

It is felt that some guidance can usefully be given to skippers of fishing vessels with regard to reasonable assumptions of ice formation on which stability conditions can be based.

A wealth of experience of this hazard has been gained by Russia whose merchant ships generally have to contend with it as a matter of routine. The following recommendations are primarily the result of this experience.

In considering the stability in conditions of ice accumulation the displacement should be corrected for the weight of ice. This weight of ice should not be included as a deadweight item in any stated condition but considered as a weather hazard in the same way as wind and shipped water, i.e. an overload. Assumptions as to weight of ice involve consideration of surface areas of decks and superstructure and of fittings. The effect of this weight involves the height of its centre of gravity.

In selecting a ship condition for investigation of ice accumulation, the one selected should be the worst from a stability point of view. As to what might be acceptable as a standard for fishing vessels in general, one would have to see that the iced-up condition did not fall below minimum stability standards as discussed in Chapter 5.

Quantity of horizontal ice accumulation should be assumed at the rate of not less than 6 lb./sq. ft.* for all exposed surfaces of weather decks, house tops and gangways. The projections of deck machinery and fittings can be ignored in the horizontal plane.

Quantity of vertical ice accumulation should be assumed at the rate of not less than 5 lb./sq. ft.* for the ship's projected lateral plane (area) above the *LWL*. The area of discontinuous surfaces (rails, spars, rigging, etc.) should be included by adding 5% to the projected lateral area of the continuous surfaces (and static moments by 10%).

The height of the centre of gravity of the ice has to be estimated or calculated by taking moments about some convenient reference point. These assumptions of ice formation apply to latitudes north of 66° 30′ N or south of 60° 00′ S, and the Barents, Bering and Okhotsk Seas, the Tatar Strait and Canadian East Coast in winter. In other areas of the winter seasonal zone assumed ice accumulations may be taken at one-half the above figures.

---

*The Canadian Steamship Inspection (C.S.I.) Service requires rates of not less than 11 lb./sq. ft. and 7.5 lb./sq. ft. for horizontal and vertical ice accumulation respectively.

Stability Information
Aboard Ships

OFFICIAL regulations relating to the carriage of stability information now apply to fishing vessels of 12 metres in length or over. Builders of fishing vessels are now required by most Administrations to supply detailed information for the guidance of the skipper and base this on an inclining test performed on all new designs.

This information takes a generally standardised form and it may sometimes be very full in extent. Some of the information, if extensive, may be somewhat beyond the comprehension of the average seagoing officer but this is not necessarily superfluous because the information does provide a basis for expert opinion on the vessel wherever she may be if the skipper requires it.

The usual information supplied may be expected to comprise:

1. A General Arrangement Plan of the ship.

2. A General Arrangement Plan of the Machinery Space.

3. A Rigging Plan (important for fishing vessels, but not always held aboard every ship).

4. A Capacity Plan showing the capacity and centres of each compartment and containing a deadweight/displacement scale and freeboard (and loadlines for ships other than fishing vessels).

5. Copy of the Inclining Test Report showing how the Light-ship particulars have been derived.

6. Instructions on the proper use of any anti-rolling devices.

7. Hydrostatic curves or Tables.

8. Cross Curves of Stability.

9. Conditions of Loading.

The last three are put together in what is usually called a "Trim and Stability" booklet which must be kept on board at all times in the custody of the Skipper. Besides a set of hydrostatic curves and cross curves of stability the booklet will contain various "Conditions of Loading" which, for a fishing vessel should comprise:

a. Absolute Lightship (builders' condition-ship completely empty but any permanent ballast specified).

b. Working Lightship (all fishing gear aboard and perhaps crew and effects, but all this will be specified)

c. Departure from port (as b) plus all fuel, water, stores, ice etc.).

d. Arrival at fishing grounds.

e. Fishing grounds half-trip condition.

f. Departure from fishing grounds (full catch).

g. Arrival in port (full catch + 10% fuel and stores).

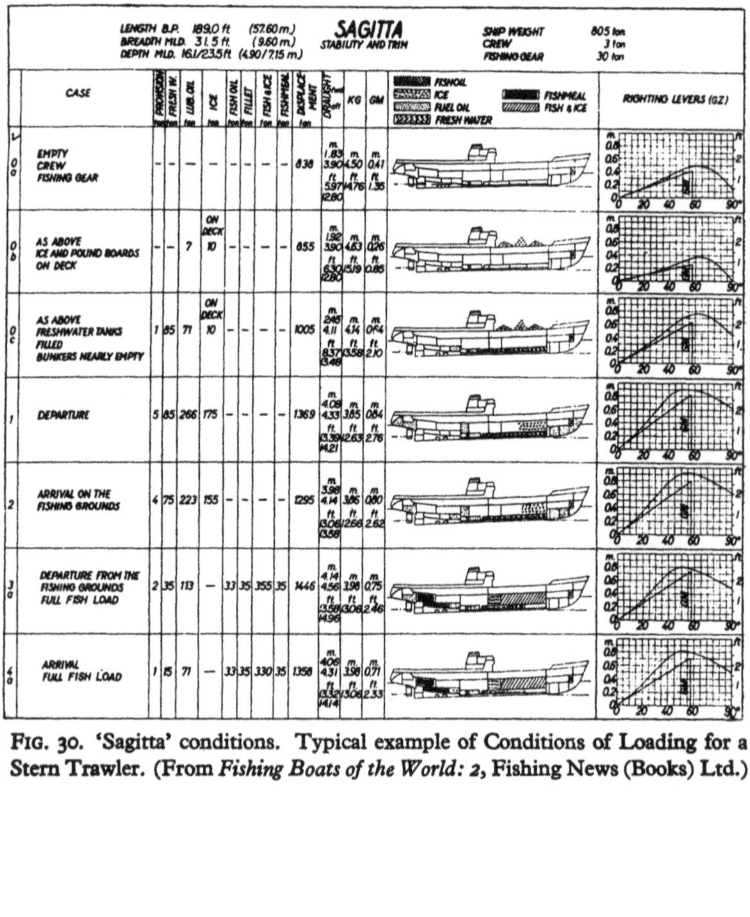

FIG. 30. 'Sagitta' conditions. Typical example of Conditions of Loading for a Stern Trawler. (From *Fishing Boats of the World: 2*, Fishing News (Books) Ltd.)

Somewhere around conditions (e.) to (f.) it may be required to give an intermediate worst stability condition with a heavy catch on deck with another load on the derrick (suspended weight, see Chapter 2, Fig. 5). For Arctic waters, an iced-up condition is also required.

Each condition of loading must give a tabular statement of all the deadweight items, a corresponding displacement, *VCG, GM* both solid and corrected for slack tanks (free surface), the drafts (mean, forward and aft), freeboard and the trim by the stern. It should also give the height of the transverse metacentre *KM*. Each condition must be accompanied by a statical stability curve corrected for free surface effects as well.

Sometimes the booklet (especially for large fishing vessels and fish factories) will give other supplementary information, e.g. effect of trim on metacentric height, increase in displacement for trim and approximate changes in draft due to filling tanks or adding specified weights in various holds or cargo compartments. An important table which must always be included is that giving the loss of *GM* due to free surface in slack tanks.

It almost goes without saying that no condition should be shown in which the stability is unsatisfactory unless there is a good reason and then only with a prominent note of warning.

Many small ships and fishing vessels incorporate a small amount of permanent solid ballast which is, or should be, part of the ship-builders' Lightship weight (it is *not* a deadweight item). The operative word is *permanent*. The inherent liability of such ballast is the possibility of its removal or part-removal without proper authorisation or control, usually to effect repairs, after which replacement is forgotten or mismanaged. It is necessary, therefore, to lay down some rules. The nature of the ballast must first of all be acceptable. For instance, an aggregate of cement and small steel scrap or pig iron kentledge is suitable whereas stone rubble ballast collected off a beach and just dumped in the bottom of the boat as the author has seen done in some countries is not (though it may be suitable as temporary ballast in other circumstances). The extent or distribution and weight of the permanent ballast must be known which involves itemising individual weights with their location in a table of vertical and longitudinal (horizontal) centres of gravity and moments in the conventional way; and this

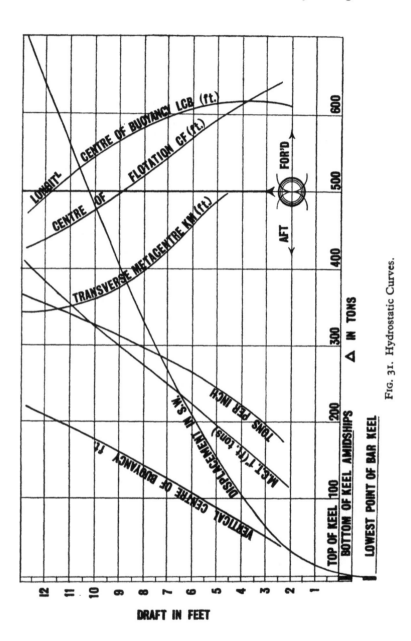

FIG. 31. Hydrostatic Curves.

applies whether the ballast is portable (slabs) or not. The solid ballast can then be checked at any time. Last but not least, whatever the form or nature of the ballast it must be arranged so that it cannot shift at sea. This is one of the reasons for having the ballast of a nature which makes it difficult for it to do this in the first place. The ocean floor is littered with the consequences of shipmasters who either did not know or did know and did not care.

Stability information requires up-dating if the ship undergoes structural modification, conversion or alteration of ballast or heavy tackle or gear. Indeed, after any major conversion the ship should be inclined afresh. The owners' superintendent or the skipper have a responsibility in this connection, it is felt, as well as the repairers. In any case, most Administrations now require to sight and approve the form and content of stability information carried aboard ship.

HYDROSTATIC CURVES

This is a most important document which shows all the elements entering into a ship's stability and trim which are solely dependent upon geometrical considerations and the shape of the underwater hull form. If these particulars are not depicted in the form of curves, a tabular statement will be supplied instead. Sometimes both are supplied.

The curves are drawn on a graph the base of which sets out varying displacements in tons. The vertical axis gives the draft in feet from some reference point which is either the top or the bottom of the keel (i.e. either moulded or extreme draft). The following curves are customarily shown:

Displacement: This is nearly a straight line drawn from the origin. It applies to seawater unless two curves are given one of which would apply to freshwater (in which case they will be marked).

Vertical Centres of Buoyancy (VCB): Also nearly a straight line drawn from the origin. The horizontal measure from the vertical gives the height value at any given draft (to a scale noted on the curve).

Transverse Metacentres (KM): The horizontal measure from the vertical axis gives the height value at any given draft (to the scale noted). Similar remarks apply to the curve of Longitudinal Metacentres if given.

Tons per Inch Immersion (TPI):* Addition or removal of weight will give bodily sinkage or rise (i.e. change in mean draft) of 1″ at any given draft. Measure horizontally to vertical axis (to scale noted).

Moment to Change Trim 1″ (MCT 1″):* The moment to effect this at any particular draft is given in tons ft. Measure horizontally to vertical axis (to scale noted).

The remaining curves have to be expressed in relation to their position to amidships. A vertical reference line is therefore drawn at any convenient position on the graph and denoted 'amidships'. The parts of curves to the left of this line are values aft of amidships and to the right forward of amidships. Such curves are used in trim problems.

Longitudinal Centres of Buoyancy (LCB): Measure to amidships reference line for draft considered and obtain value, according to scale given, forward or aft of amidships.

Centres of Flotation (CF): Measure to amidships reference line for draft considered. This gives the *CF* or tipping centre of the waterplane either forward or aft of amidships (to scale noted).

Most merchant ships are designed 'on even keel'. Many fishing vessels, however, have a designed trim or keel drag and where this is so the lowest point of the keel (aft) will not be the same as the bottom (or underside) of keel amidships. In this case, the horizontal baseline of the graph will be most unlikely to represent this latter point (for reasons to do with the way in which the designer does the hydrostatic calculations). This makes the reading of extreme drafts direct from the vertical scale incorrect unless the amount of keel rake is added. This will, however, be given on the graph.

It is also worth mentioning that the horizontal separation (difference) of the curves of *VCB* and *KM* at any draft equals the height of *KM* above the *VCB*, i.e.

$$BM \left(\text{or } \frac{I}{V} \right) = KM - KB$$

---

*Metric values are in Tonnes per Centimetre Immersion and Moment to Change Trim One Centimetre respectively.

FORMULAE AND APPROXIMATIONS

RADIAN—(Scientific unit of measurement for angles). The angle subtended at the centre of any circle by an arc equal in length to its radius. Thus:

2π radians correspond to 360 degrees

1 radian corresponds to $\dfrac{180}{\pi} = 57.3$ degrees

$\dfrac{\pi}{180}$ radians correspond to 1 degree

Length of arc = radius × angle in radians, i.e. $1 = r\theta$

Moment of Statical Stability $= \Delta GZ = \Delta GM \sin \theta$

$$BM = \frac{\text{Moment of Inertia of Waterplane}}{\text{Volume of Displacement}} = \frac{I}{V}$$

Approximate $BM = \dfrac{a \times B^2}{\text{draft} \times Cb}$, where B is breadth of ship or

waterplane, Cb is the block coefficient and a is a constant depending on coefficient of waterplane area according to table below

Cw	0.5	0.6	0.7	0.8
a	0.021	0.031	0.042	0.055

Approximations for Cw used are:

$Cw = Cb + 0.10$ or $Cw = \sqrt{Cb} - 0.025$

VCB below $LWL = \dfrac{1}{3}\left\{\dfrac{d}{2} + \dfrac{\nabla}{A}\right\}$ (Morrish's formula)

where d = mean draft and $A = WP$ area.

$$KB = \frac{1}{6}\left\{5d - \frac{2\nabla}{A}\right\} \qquad KM = 0.555d + \frac{.0777\,B^2}{d}$$

Inclining Test—— $GM = \dfrac{w \times d}{\Delta\,\tan\theta}$, where d = shift.

$\tan\theta = \dfrac{\delta}{l}$, where δ = deflection of pendulum and l is its length.

$$\therefore GM = \frac{w \times d \times l}{\Delta \times \delta}$$

Free Surface Effect—Loss of $GM = \dfrac{i}{\nabla}\left\{\text{or } \dfrac{i}{\Delta \times 35}\right\}$

But if liquid in ship not seawater correct for density,

i.e. $\dfrac{\text{density of liquid}}{\text{density of sea}} \times \dfrac{i}{\nabla}$

Large Angle Stability

$$GZ = \frac{v \times hh_1}{\nabla} - BG\,\sin\theta$$

where $v \times hh_1$ is transfer moment of wedge volume horizontally.

Atwood's Formula:

Moment of Statical Stability $= \Delta\left\{\dfrac{v \times hh_1}{\nabla} - BG\,\sin\theta\right\}$

If vessel 'wall-sided' in region of water line then,

$$GZ = \sin\theta\left\{GM + \frac{1}{2}BM\,\tan^2\theta\right\}$$

If GM negative, then $GZ = 0$ (zero)

and $\tan\theta = \pm\sqrt{2\,\dfrac{GM_1}{BM_1}}$

$\therefore GM$ at 'Angle of Loll' $= 2\,\dfrac{GM_1}{\cos\theta}$

Correction to Statical Stability Curves levers (GZ's) for alteration in KG (or for other than assumed position)

$+ GG_1 \sin \theta$ (if G_1 below G)

$- GG_1 \sin \theta$ (if G_1 above G)

$$MCT\ 1'' = w \times d = \frac{\Delta\ GM_L}{12L}\ \text{tons ft.}$$

or $MCT\ 1\text{cm} = \dfrac{\Delta\ GM_L}{100L}\ \text{tonne metres}$

Approximations:

$$MCT\ 1'' = \frac{\Delta}{10}\ \text{or}\ \frac{L\ \Delta}{\text{draft} \times 190}\ \text{or}\ 0.000175\ \frac{A^2}{B}$$

$$MCT\ 1\text{cm} = \frac{7(TPCm)^2}{B}\ \text{tonne metres}$$

$$\text{Tons per Inch Immersion} = \frac{\text{Area of } WP}{420}$$

$$\text{Tonnes per Centimetre Immersion} = \frac{A}{100} \times 1.025$$

$$\text{Change in draft due to change in density} = \frac{\Delta\ (\delta_1 - \delta_2)}{\delta_2 \times TPI}\ \text{inches}$$

$$\text{Ship's natural period of roll } 2T = \frac{1.108\ k}{\sqrt{GM}}$$

$$GM = \left\{\frac{1.108\ k}{2\ T}\right\}^2$$

$$\text{or} = \left\{\frac{1.108\ mB}{2\ T}\right\}^2$$

(The numerator $1.108\ mB$ may be incorporated into a factor F which applies to a particular ship at a particular condition of loading.)

METRIC CONVERSIONS and EQUIVALENTS

Relation between Weight and Volume:

```
10 mm. cubed  =  1 cu. cm.
 1 cu. metre  =  35.316 cu. ft.
 1 cu. ft.    =  0.0283 cu. metre
 1 cu. cm.    of freshwater (SG = 1.0)   = 1 gramme
1000 cu. cm. ,,      ,,          ,,      = 1 kilogram (1000 gm.)
 1 cu. metre ,,      ,,          ,,      = 1 tonne (1000 kilos)
 1 cu. metre ,, saltwater (SG = 1.025)  = 1.025 tonnes
 1 tonne     ,,      ,,          ,,      = 0.975 cu. metres
```

MULTIPLY BY	TO CONVERT FROM	TO OBTAIN	—
0.03937	Millimetres	Inches	25.400
0.3937	Centimetres	Inches	2.540
3.2808	Metres	Feet	0.3048
2.2046	Kilogrammes	Pounds (lb.)	0.45359
0.0009842	Kilogrammes	Tons (2240 lb.)	1016.0470
0.9842	Metric Tonnes (1000 Kilos)	Tons (2240 lb.)	1.0160
2.4998	Metric Tonnes per Cm. Immersion	Tons per Inch Immersion	0.400
8.2014	M.C.T. 1 Cm. (Tonne Metre units)	M.C.T. One Inch (Ton Ft. units)	0.122
187.9767	Metre Radians	Feet Degrees	0.0053
—	TO OBTAIN	TO CONVERT FROM	MULTIPLY BY ABOVE

Appendix I

Change in Draft due to Change in Density

δ_f and δ_s are the densities in oz./cu. ft.

Δ is displacement in tons; ∇ is volume in cu. ft.

$$\Delta = \frac{\nabla_f \times \delta_f}{16 \times 2240} \qquad \begin{cases} 16 \text{ oz.} = 1 \text{ lb.} \\ 2240 \text{ lb.} = 1 \text{ ton} \end{cases}$$

$$\therefore \quad \nabla_f = \frac{16 \times 2240 \times \Delta}{\delta_f}$$

$$\text{and} \quad \nabla_s = \frac{16 \times 2240 \times \Delta}{\delta_s}$$

Subtract $\nabla_f - \nabla_s = 16 \times 2240 \times \Delta\left\{\frac{1}{\delta_f} - \frac{1}{\delta_s}\right\}$

$$= 16 \times 2240 \times \Delta\left\{\frac{\delta_s - \delta_f}{\delta_f \, \delta_s}\right\}$$

$\nabla_f - \nabla_s = $ change in draft (ft.) $\times$ Area of WP

$$\therefore \text{Change in draft (ins.)} = \frac{\nabla_f - \nabla_s}{A} \times 12$$

$$= \frac{12 \times 16 \times 2240 \times \Delta\,(\delta_s - \delta_f)}{A \times \delta_f \, \delta_s} \qquad . \quad (1)$$

But

$$TPI = \frac{A}{12} \times \frac{\delta_s}{16 \times 2240}$$

$$\therefore A = \frac{TPI \times 12 \times 16 \times 2240}{\delta_s} \qquad . \quad . \quad . \quad . \quad . \quad (2)$$

Substituting for A in (1)

$$\text{Change in draft (ins.)} = \frac{\Delta\,(\delta_s - \delta_f)}{TPI \times \delta_f}$$

(or change in draft (cm.) if Δ in tonnes and using $TPCm$.)

Appendix 2

Moment of Inertia of a Waterplane about Ship's Fore-and-Aft Centreline

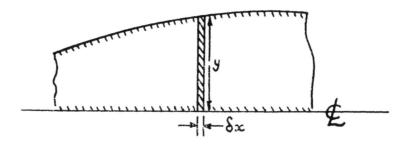

Consider an elementary strip y of thickness δx. This may be assumed rectangular. The inertia i of a rectangle about an axis passing through one side is $\dfrac{\text{breadth} \times \text{depth}^3}{3}$

In this case $i_{\text{\pounds}} = \dfrac{y^3 \, \delta x}{3}$

For both sides $i_{\text{\pounds}} = \dfrac{2}{3} y^3 \, \delta x$

For whole waterplane $I_{\text{\pounds}} = \dfrac{2}{3} \Sigma y^3 \, \delta x$

i.e. $\underset{\delta x \to 0}{Lt} \ \dfrac{2}{3} \Sigma y^3 \delta x = \dfrac{2}{3} \int y^3 dx$

Using Simpson's Rules and $\frac{1}{2}$ ordinates:

$$I = \frac{2}{3} \times \frac{1}{3}^{*} \times C.I. \times \Sigma fn\, I$$

** or $\dfrac{3}{8}$

If $I_{\mathcal{L}}$ is required for an off-centreline area then this is found by application of the Theorem of Parallel Axes, i.e. $I_{\mathcal{L}} = I_{xx} + Ad^2$

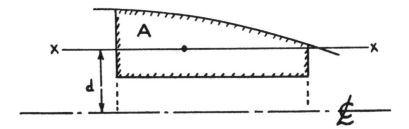

where A is area and xx is axis through centroid of area.

Appendix 3

Formula $BM = \dfrac{I}{V}$

"The distance between the Centre of Buoyancy and the Transverse Metacentre equals the Moment of Inertia of the Waterplane about the Centreline divided by the Underwater Volume of the Ship."

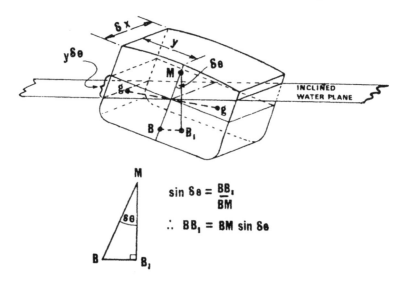

$$\sin \delta\theta = \frac{BB_1}{BM}$$

$$\therefore \; BB_1 = BM \sin \delta\theta$$

The above figure represents a small section of a ship of width δx which is heeled through a small angle $\delta\theta$. The centre of buoyancy shifts from B to B_1. Since $\delta\theta$ is small the vertical through B_1 cuts the centreline of the body at M, the transverse metacentre.

But as $\delta\theta$ small, $\sin \delta\theta$ approximately equals $\delta\theta$.

$$\text{Hence } BB_1 = BM\delta\theta$$

As vessel heels a wedge of buoyancy with its centre of gravity at g (the Emerged Wedge) is transferred to the other side of the ship with centre of gravity at g_1 (the Immersed Wedge). Their volumes are equal at small angles of inclination. Let v be the volume of a wedge.

Then $\nabla \times BB_1 = v \times gg_1$ (and BB_1 is parallel to gg_1)

Substituting for BB_1

$\nabla \times BM\delta\theta = v \times gg_1$

But $v = \frac{1}{2} b \times h \times l$ (properties of a triangle)

i.e. $= \Sigma \frac{y}{2} \times y\delta\theta \times \delta x$

$$= \Sigma \frac{y^2 \, \delta\theta \, \delta x}{2}$$

And $gg_1 = \frac{4}{3} y$ (centre of gravity of a triangle)

$\therefore v \times gg_1 = \Sigma \frac{y^2 \, \delta\theta \, \delta x}{2} \times \frac{4}{3} y$

$$= \frac{2}{3} \Sigma y^3 \, \delta x \, \delta\theta$$

But $\frac{2}{3} \Sigma y^3 \, \delta x = I_{\mathfrak{L}}$ (Appendix 2)

Then $v \times gg_1 = I_{\mathfrak{L}} \, \delta\theta$

$\therefore \nabla \times BM \, \delta\theta = I_{\mathfrak{L}} \, \delta\theta$

$\therefore \qquad BM = \frac{I}{V}$

(N.B. V is frequently written ∇ to distinguish it from small v).

Note:—This formula is of universal validity in its application to any inclination. Thus, not only does it apply as above but also to longitudinal metacentres coupled with longitudinal moments of inertia and any combination thereof involving waterplanes inclined due to simultaneous heel and trim.

Appendix 4

Moment to Change Trim 1 in. (or 1 cm.)

This is usually abbreviated $MCT\ 1''$

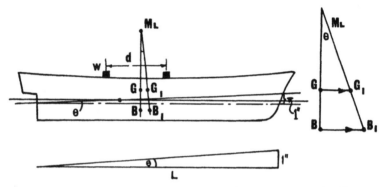

It is required that a trimming moment of weight w moved a distance d (i.e. $w \times d$ tons ft.) shall trim the ship $1''$.

G moves to G_1

Taking moments: $w \times d = \Delta \times GG_1$ . . . . (1)

Also $\tan \theta = \dfrac{GG_1}{GM_L}$

θ small so $\tan \theta \approx \theta$

$\therefore GG_1 = GM_L\ \theta$ . . . . (2)

Substituting (2) into (1)

$w \times d = \Delta GM_L\theta$ . . . . (3)

But $\theta = \dfrac{\text{Trim in feet}}{L}$

$= \dfrac{1/12}{L} = \dfrac{1}{12\ L}$

Substituting into (3)

$$w \times d \quad = \frac{\Delta\, GM_L}{12\, L}$$

i.e. $MCT\ 1'' = \dfrac{\Delta\, GM_L}{12\, L}$ or $MCT\ 1\ \text{cm.} = \dfrac{\Delta\, GM_L}{100\, L}$

Where GM_L is called the *Longitudinal Metacentric Height*

And M_L is called the *Longitudinal Metacentre*.

N.B. If metric, Δ must be in tonnes and L in metres.

Appendix 5

Displacement Correction for Trim

The mean draft (at midships) of a ship having trim does not correspond to the actual displacement as given by a Displacement Curve. As the the ship trims about the CF of the original waterplane, the draft corresponding to the displacement occurs at this CF. Hence to the displacement at mean draft amidships must be added or subtracted a 'layer correction'. (See Fig. 18.)

If θ is a small angle of trim

Then $\tan \theta = \dfrac{\text{Trim}}{L}$ (where L is length of ship)

(Trim is amount out of designed trim if designed trim is not even keel.)

but $a - b$ is approx. thickness of layer (t)

Also $\tan \theta = \dfrac{\text{thickness of layer } (t)}{\text{distance of } CF \text{ from midships } (y)}$

$\therefore \dfrac{\text{Trim}}{L} = \dfrac{t}{y}$

or $t = \dfrac{y}{L} \times \text{Trim}$

but $\delta\Delta = TPI \times t$ (where $\delta\Delta =$ small increase in displacement)

$\qquad\quad = TPI \times \dfrac{y}{L} \times \text{Trim}$

The increase in displacement must be added or deducted according to whether the CF is forward or abaft of midships and by the head or by the stern (i.e. 4 alternatives). The above applies only to moderate trim changes for which the displacement correction, being relatively small, is not of great practical importance to seamen.

Appendix 6

International Co-operation in Stability Studies

Co-operation in stability studies for fishing vessels on an international basis is led by the International Maritime Organisation, IMO (previously Inter-governmental Maritime Consultative Organisation, IMCO) assisted by the Food and Agriculture Organisation (FAO).

As a result of comprehensive studies of existing national requirements, on analyses of intact stability casualty records and on stability calculations of ships which have operated successfully, a Recommendation on Intact Stability of Fishing Vessels was drawn up containing recommended stability criteria for fishing vessels together with a number of other related recommended practices.

By Resolution A.168(ES.IV) the IMCO Assembly invited all governments concerned to take steps to give effect to the Recommendation as soon as possible unless they were fully satisfied that their national stability requirements, supported by long operating experience, already ensure adequate stability for particular types and sizes of vessels. The Assembly at the same time requested continuation of studies of stability of fishing vessels, including the formulation of specific requirements related to the icing of vessels off the east coast of Canada during winter months, and development of improved stability criteria.

The IMO Recommendations in conjunction with those of the Food and Agricultural Organisation and the International Labour Organisation, are now enshrined in a CODE OF SAFETY FOR FISHERMEN AND FISHING VESSELS which is in two parts, viz:

Part A — Safety and Health Practice for Skippers and Crews.
Part B — Safety and Health Requirements for the Construction and Equipment of Fishing Vessels.

The Code is updated or amended as found necessary in the light of further experience and IMO acts as a focal point for co-ordination.

A major advance in international co-operation in this field was the Torremolinos International Convention for the Safety of Fishing Vessels, 1977. This was the first such International Convention covering safety requirements for the construction and equipment of fishing vessels.

Appendix 7

Suggestions for Preserving Safe Stability

The following measures are recommended by IMO with some amplification in places by the author. They should be considered as guidance on matters influencing the safety of fishing vessels generally, and specifically as related to safeguarding stability.

1. All doorways and other openings through which water can enter the hull or deckhouses, forecastle, etc. should be suitably closed in adverse weather. Accordingly, all appliances for this purpose should be maintained aboard in good and efficient condition. Enclosed spaces like deckhouses above the weather deck contribute to stability but if their doors or openings are forced open to the sea and water collects inside this may be slow to drain away. In this case, the flooding could become dangerous to the safety of the vessel.

2. Hatch covers and flush deck scuttles should be kept properly secured when not in use during fishing.

3. All deadlights should be maintained in good condition and securely closed in bad weather.

4. All fishing gear and other large weights should be properly stowed and placed as low as possible.

5. Care should be taken when the pull from fishing gear might have an adverse effect on stability, e.g. when nets are hauled by power block or the trawl catches obstructions on the sea bed. This is particularly the case when the vessel is manoeuvring with the trawl abeam and is the worst for small vessels. Gallows frames are the same size in general for small vessels as for the larger ships which means for the former a relatively greater lever. The point of action of the weight is at the hoist block of the frame or derrick head (see Chapter 2 "Suspended Weights").

6. Gear for releasing the deck load in fishing vessels carrying catch on deck, e.g. herring, should be kept in good working order for immediate use when necessary.

7. Freeing ports in bulwarks which are provided with closing appliances should always be capable of functioning and should not be locked, especially in bad weather. Devices for locking freeing port covers should be regarded as potentially dangerous. If locking devices in particular cases are considered essential for

the service of the ship, they should be of a reliable type, operative from a position which would always be accessible. When operating in areas subject to ice formation, it is recommended not to fit covers at all. Water on deck in the well between bridge and forecastle or elsewhere can be a hazard to stability unless cleared rapidly. Moreover, this can build-up by an equal amount with the onset of each successive wave.

8. When the weather deck is prepared for the carriage of deck load by division with pound boards, there should be slots between them of a size such that an easy flow of water to the freeing ports will be ensured, i.e. good drainage.

9. Never carry fish in bulk without first being sure that the portable divisions in the fish hold are properly installed. The cargo must not shift.

10. At any one time keep the number of partially filled tanks to a minimum.

11. Observe any instructions given regarding the filling of water ballast tanks. Remember that slack tanks can be dangerous.

12. Any closing devices provided for vents to fuel tanks etc. should be secured in bad weather.

13. Reliance on automatic or fixed steering is dangerous as this prevents speedy manoeuvring which may be needed in bad weather.

14. Be alert to all the dangers of following or quartering seas. These may cause heavy rolling and/or difficult steering. If excessive heeling or yawing occurs, reduce speed or alter course or both.

15. Maintain a seaworthy freeboard in all conditions of loading. Remember that this has a very marked effect on the vessel's maximum righting and recovery powers and the range of heeling angles over which the ability to recover depends.

16. Pay special attention to the formation of any ice aboard the vessel and reduce it by all possible means. Standing wire rigging will ice-up to a greater extent than struts or yards. If icing cannot be controlled leave the area with all possible speed long before it becomes a serious menace.

Appendix 8

Suggested Form of Guidance to Masters on an Approximate Determination of Ship's Stability by Means of the Rolling Period Test

Introduction

1. If the following instructions are properly carried out, this method allows a reasonably quick and accurate estimation of the metacentric height, which is a measure of the ship's stability.

2. The method depends upon the relationship between the metacentric height and the rolling period in terms of the extreme breadth of the vessel.

Test Procedure

3. The rolling period required is the time for one complete oscillation of the vessel and to ensure the most accurate results in obtaining this value the following precautions should be observed:

    a. The test should be conducted with the vessel in harbour, in smooth water with the minimum interference from wind and tide.

    b. Starting with the vessel at the extreme end of a roll to one side (say port) and the vessel about to move towards the upright, *one complete oscillation* will have been made when the vessel has moved right across to the other extreme side (i.e. starboard) and returned to the original starting point and is about to commence the next roll.

    c. By means of a stop-watch, the time should be taken for not less than about 5 of these complete oscillations; the counting of these oscillations should begin when the vessel is at the extreme end of a roll. After allowing the roll to completely fade away, this operation should be repeated at least twice more. If possible, in every case the same number of complete oscillations should be timed to establish that the readings are consistent, i.e. repeating themselves within reasonable limits. Knowing the total time for the total number of oscillations made, the mean time for one complete oscillation can be calculated.

    d. The vessel can be made to roll by rhythmically lifting up and putting down a weight as far off middle-line as possible; by pulling on the mast with a rope; by people running athwartships in unison; or by any other means. However, and this is most important, as soon as this forced rolling is sufficiently developed the means by which it has been induced must be stopped and the vessel allowed to roll freely and naturally. If rolling has been induced by lowering or raising a weight it is preferable that the weight is moved by a dockside crane. If the ship's own derrick is used, the weight should be placed on the deck, at the middle-line, as soon as the rolling is established.

e. The timing and counting of the oscillations should only begin when it is judged that the vessel is rolling freely and naturally, and only as much as is necessary accurately to count these oscillations.

f. The moorings should be slack and the vessel 'breasted off' to avoid making any contact during its rolling. To check this, and also to get some idea of the number of complete oscillations that can be reasonably counted and timed, a preliminary rolling test should be made before starting to record actual times.

g. Care should be taken to ensure that there is a reasonable clearance of water under the keel and at the sides of the vessel.

h. Weights of reasonable size which are liable to swing, (e.g. a lifeboat), or liable to move (e.g. a drum), should be secured against such movement. The free surface effects of slack tanks should be kept as small as is practicable during the test and the voyage.

Determination of Initial Stability

4. Having calculated the period for one complete oscillation, say T seconds, the metacentric height GM can be calculated from the following formula:

$$GM = \frac{F}{T^2}$$

where F is . . . (to be determined for each particular vessel by the Administration).

5. The calculated value of GM should be equal or greater than the critical value which is . . . (to be determined for each particular vessel by the Administration).

Limitations on the Use of this Method

6. A long period of roll corresponding to a GM of 8" (0.20 m.) or below, indicates a condition of low stability. However, under such circumstances, accuracy in determination of the actual value of GM is reduced.

7. If, for some reason, these rolling tests are carried out in open, deep but smooth waters, inducing the roll, for example, by putting over the helm, then the GM calculated by using the method and coefficient of paragraph 3 above should be reduced by (figure to be estimated by the Administration) to obtain the final answer.

8. The determination of stability by means of the rolling test in disturbed waters should only be regarded as a very approximate estimation. If such test is performed, care should be taken to discard readings which depart appreciably from the majority of other observations. Forced oscillations corresponding to the sea period and differing from the natural period at which the vessel seems to move should be disregarded. In order to obtain satisfactory results, it may be necessary to select intervals when the sea action is least violent, and it may be necessary to discard a considerable number of observations.

Index

Lightning Source UK Ltd.
Milton Keynes UK
UKHW041833250220
359320UK00001B/6